AMBIOPHONICS

Beyond Surround Sound
to Virtual Sonic Reality

Ambiophonics Institute
Northvale, New Jersey

AMBIOPHONICS

NOTICES

Pillobaffle
RPG Diffusor System, Inc.

Tube Traps
Shadow Caster Acoustic Sciences Corp.

First Edition
First Printing 1995
Printed in the United States of America

© 1995 by Ralph Glasgal

Printed in the United States of America, all rights reserved. The authors and publisher take no responsibility for the use of any of the materials or methods described in this book, nor for the products thereof.

Library of Congress Catalog Card Number: 95-77173

GLASGAL, Ralph
YATES, Keith

AMBIOPHONICS - Beyond Surround Sound To Virtual Sonic Reality /by Ralph Glasgal and Keith Yates - First Edition Includes Index and Bibliographical References

ISBN 0-9646634-0-6

1. Home Sound Reproduction
2. Audio
3. High Fidelity
4. Psycho Acoustics
5. Music Recording Playback

Cover photograph: George Pierce, Pompton Lakes, New Jersey

Published by Ambiophonics Institute
      151 Veterans Drive
      Northvale, NJ 07647
      201-768-8082 Ext.205
      Fax 201-768-2947

DEDICATION

To Johann Sebastian Bach, Richard Wagner, Gustav Mahler, Marilyn Horne, and Lauritz Melchior, without whom we would never have bothered, and to Manfred Schroeder, Don Keele, Bob Carver, Peter D'Antonio, Yoicho Ando, and Floyd Toole, without whose collective research we would never have succeeded.

TABLE OF CONTENTS

INTRODUCTION 1

CHAPTER 1
AMBIOPHONICS : ACHIEVING VIRTUAL
SOUND REALITY 7

CHAPTER 2
CONCERT HALL SOUND CHARACTERISTICS 15

CHAPTER 3
STEREOPHONIC SOUND FIELDS 25

CHAPTER 4
CHOOSING AND PLACING
LOUDSPEAKERS AMBIOPHONICALLY 51

CHAPTER 5
TUNING THE LISTENING ROOM
FOR AMBIOPHONICS 57

CHAPTER 6
EARLY REFLECTION AND
REVERBERATION SYNTHESIS 79

CHAPTER 7
FUTURE AMBIOPHONIC ENHANCEMENTS AND A
SURROUND SOUND ROUNDUP 93

EPILOGUE 101

APPENDIX 103

INDEX 107

INTRODUCTION

AMBIOPHONICS: RECREATING THE CONCERT HALL EXPERIENCE AT HOME

There are essentially only two ways for music lovers to enjoy music performed for them on traditional acoustic instruments. One is by going to a concert hall or other auditorium, and the second is by staying home and playing the radio/TV or a recording. This book and the techniques it describes are dedicated to helping you make the home music-listening experience as audibly exciting as the live experience. Those audiophiles who share the dream of recreating a concert-hall sound field in their home, and who constantly strive to create a sense of "you-are-there," We have christened "ambiophiles". We call the science and technology used to create such an acoustic illusion "ambiophonics".

DEFINING THE PROBLEM

Barry Willis wrote in Stereophile Magazine (August, 1994), "The idea that any musical event can be reproduced accurately through a two-channel home-audio system in a room that in no way resembles the space in which the original event took place is ludicrous."

Mr. Willis was absolutely correct in this when he wrote those words, but is much less so now, because ambiophonics successfully works and its purpose is precisely to make the home-audio system room resemble the space in which the original event took place. He goes on, "At present, even the best discrete multichannel surround systems can offer only an illusion of being there." Experienced ambiophiles (a rare breed) would agree, but would also point out that surround sound is deliberately designed to produce the illusion of "they-are-here-around-you" which, while exciting, is always going to be the antithesis of "being there". Finally, thoroughly despondent, Mr. Willis writes, "Tonal accuracy is the best that can be hoped for in a traditional audio system; true spatial accuracy will never happen. Audio products should come bearing this disclaimer: WARNING: IMAGE PRESENTED IS LESS THAN LIFELIKE." The rest of us need not despair. Seven years of experiments have been devoted to demonstrating that "lifelike" can happen, and with exceptional fidelity to the original, from just two standard stereo channels. Yes, the ambiophonic method described below may not always duplicate a particular hall with precision, but it can create a hall that could exist architecturally, that rings true, and is lifelike enough to mimic a good seat at a live musical event.

TRADITIONAL AUDIOPHILE ARTICLE OF FAITH

Many, if not most, serious audio enthusiasts presently believe that it is possible to achieve a solid stage image that may even extend beyond the loudspeaker positions, by employing the usual arrangement where two loudspeakers and the listener form something close to an equilateral triangle. They have faith that the perfect loudspeaker, amplifier, CD or record player and associated cables will produce that wide, sharp imaging, stage depth, and ambient clarity that we all prize. Many also believe that audiophile-grade equipment, properly selected and tweaked, combined with signal path minimalism, is more likely than simple acoustic listening-room treatments to produce a higher fidelity sound field with enhanced width and depth. Some audio hobbyists prefer to stay with a small ensemble "they-are-here" jazz-combo sound and have no need, or even desire, to achieve a realistic orchestral or operatic sound. They feel strongly that this is not what high-end sound reproduction should concern itself with. Many experienced listeners also hold that the hall reverberation captured by the recording microphones is being properly reproduced when it comes, together with the direct sounds, from the front loudspeakers.

A new breed of video-age audiophile is convinced that hall ambience, extracted from specially encoded recordings and steered to surround speakers by surround-sound decoders can achieve the "you-are-there" illusion. This latter group is at odds with those who hold that any such processing, especially in the analog signal chain, is anathema.

Considering these prevailing and conflicting conceptions and misconceptions, it is remarkable how good, and even exciting, a sound can be produced by such high-fidelity equipment and methods. The musical sound generated by products from the overwhelming majority of acknowledged high-end equipment manufacturers is truly first class. But the traditional methods of deploying this superb equipment at home has reached a dead end as far as closing that last yawning gap between perfect, but flat, fidelity and true spatial realism.

AMBIOPHONICS—THE NEXT AUDIOPHILE PARADIGM

We believe that the majority of serious home music listeners are closet ambiophiles who really want to be in a realistic, electronically created concert hall, church, jazz club or opera house when listening to recorded music at home. The purpose of this book is to pass on the results of the research and experiments that we have performed. Ambiophiles everywhere can take comfort in the fact that it is both theoretically possible, possible in practice, and reasonable in cost to achieve the formerly impossible dream of synthesizing a "you-are-there" soundfield from standard unencoded stereo records or CDs in virtually any properly treated room at home. In ambiophonic parlance, when we say "real," we mean that an acoustic space of appropriate size and liveness has been created that is realistic enough to fool the human ear-brain system into believing that it is within that space with the performers.

The ambiophonic techniques described in the following chapters produce a sound stage as wide as that seen by the stereo recording microphones, an early-reflection sound pattern that defines the hall size and the listener's position within the hall, and finally a reverberant field that complements the content of the music and the original recording venue and defines the character of the performing space.

Although ambiophonics does not rely on decoders, matrices or ambience extraction, it does incorporate commercially available digital signal processors, which are essentially special-purpose computers, to generate the appropriate ambience signals. It is, therefore, a prime article of the ambiophile faith that while such signal generators are always subject to improvement, they have already reached an audiophile level of performance, if one uses them ambiophonically as described in the chapters that follow. It is also not the belief of the authors that there is only one fixed way to achieve the ambiophonic result. But we believe the ambiophonic methods and principles we espouse can form a better foundation to build on than traditional stereo or surround-sound technologies.

In brief, ambiophonics uses room treatment, radical front-channel loudspeaker positioning, electronic generation of early reflections and the later reverberant field and additional loudspeakers strategically placed to accurately propagate the direct and ambient fields. Not every audiophile will be able or willing to do all that we suggest, but as each feature of the ambiophonic system is implemented the improvement in realism will be easily audible and clearly rewarding. The treated ambiophonics listening room can also be an excellent space for live music at home.

If any sciences can be called ancient, acoustics is certainly one of them. The literature on acoustics, concert-hall design and sound recording is so vast that we are prepared to concede in advance that no individual fact or idea in the chapters below has not already appeared, at some time, in print in one journal or another. We can only hope that the concatenation of all these ideas and inventions that define ambiophonics has some modicum of novelty. While we don't need to credit pioneers as far back as Helmholtz and Edison, we would like to acknowledge our debt to such relatively recent researchers as W.B. Snow, James Moir, Manfred R. Schroeder, and particularly to his former colleague Yoicho Ando, whose ideas on how to build better public concert halls inspired us to adapt his methods to create virtual halls for home concerts.

CHAPTER 1
AMBIOPHONICS: ACHIEVING VIRTUAL SOUND REALITY

Ambiophonics is a technical methodology which, if adhered to closely, makes it possible to immerse oneself in an exceedingly real acoustic space, sharing it with the music performers. Ambiophonics does this at home using standard recordings. Ambiophonics is for serious listeners who do not expect to read, knit, talk or sleep in their home concert halls, any more than they would at a live performance. We have also observed a strong tendency of ambiophiles to listen at normal concert hall volume, which does rivet one's attention.

THE AMBIOPHONIC PLAYBACK SYSTEM

Ambiophonics was developed to provide audiophiles, record collectors, equipment manufacturers and, eventually, record producers with a clear, understandable recipe for generating realistic sound fields consistently and repeatedly from standard non-surround-sound commercial recordings.

The basic elements required in the home are the following:
1. A dedicated Listening Room. A room dedicated to this purpose where decor and all other considerations are kept subservient to the requirements of the ambiophonic method and the laws of acoustics. If the growth

of home video-theatre installations is any indication, there are thousands of home videophiles who are prepared to invest in video projectors, theatre seats, large screens and built-in surround-sound systems in order to duplicate the movie-theatre experience at home. Perhaps, there are similar numbers of music lovers who are prepared to invest in a home concert-hall or opera house. Duplicating the concert hall experience at home is not nearly as expensive or complex as home theatre, but does require a similar dedicated room and a determined mind set.

2. Listening Room Treatment. While the size and shape of the room are not critical, proper sound treatment is essential. The room must be so fashioned that sound reflections from its walls are minimized and do not interfere with the illusion that ambiophonics creates. Keeping exterior noise out of the room is also a function of the room treatments discussed in the chapters following.

3. Barrier Wall. The front main left and right loudspeakers' sounds must be kept acoustically isolated at the listening position. This entails fixing the listening position and standing a permanent or folding panel on edge, extending from this seat to the space between the speakers (see chapters 3 and 5).

4. Front Speaker Placements. The front speakers must be moved to a position almost directly in front of the listener, but still separated by the barrier wall.

5. Front Proscenium Reflections. Left- and-right proscenium early reflection signals must be synthesized by computer or digital signal processors and reproduced ideally through a pair of loudspeakers placed at 55 degrees to the left and 55 degrees to the right of the lis-

tening position.

6. Side Hall Reverberation. Left- and right-side reverberant and early reflection signals must be synthesized and reproduced through loudspeakers placed to the left and right of the listening position.

7. Rear Hall Reverberation. Left and right rear-hall reverberation signals must be synthesized and reproduced by two or four loudspeakers elevated behind, or behind and above, the listening position.

8. Amplifier Power. Enough amplifier power must be available to achieve concert-hall volume at the listening position.

The technical reasons for these requirements are discussed in the chapters that follow. In some cases, once the physics or the psychoacoustic laws are understood, the reader may be able to think of other, better ways to achieve the same end.

Ambiophonics was not developed in a day, and the reader may not want to implement the entire ambiophonic system at one time. If so, we would suggest doing the barrier first, but each element in the system does result in an appreciable, audible improvement.

WHAT AMBIOPHONICS SPECIFICALLY ACHIEVES

If you employ the techniques described in the chapters following, you will produce a rock-solid sound stage that consistently extends far beyond the right and left positions of the front loudspeakers. You will find that even with the main left and right loudspeakers directly in front of you, there is not only no compromise in the perceived stage width or depth, but an improvement. You will also see that software-generated hall ambience and reverberation, if propagated in a properly treated room, launched from the correct direction by a well situated loudspeaker, and reasonably matched to the music and the recording, will yield the sense that you are in a hall similar to that in which the recording was made.

BASICS OF CONCERT PSYCHOACOUSTICS

In order to produce a concert-hall sound field in the home without actually building a concert hall, we need to know what the ear requires at the minimum for accepting a sound field as real. Knowing this, it is then possible to look for ways to accomplish it in a small space and within a budget, without compromising the aural illusion. While not everything is known about how the ear perceives horizontal and vertical angular position, enclosure size and shape, and absolute polarity, enough is known to allow ambiophonics to create a variety of sound fields suited to different types of music that are real enough to be accepted as such by the ear-brain system.

In general the only parts of the hearing mechanism that concern us specifically are the ear pinnae and the existence of two ears separated by a head. Even without consulting the hundreds of papers on this subject, it is clear that the pinnae are designed to modify the frequency response of sound waves as a function of the direction from which the sound comes. It is also clear that no two individuals have ear pinnae identically shaped. But to give a general idea of what the pinna does in the horizontal plane: for a sound coming from directly in front, the frequency response at the ear canal entrance is essentially flat up to 1000 Hertz. The response then rises, as the rear of the pinna interdicts sound and reflects it additively into the ear canal. A broad 11db peak in the response is reached at about 3000 Hz, after which the response drops off to minus 10 db at 10 kHz and then begins to rise again. A response spread such as this of 21 db in the treble region is quite substantial, and if a loudspeaker had this kind of response it would get very poor reviews indeed. It is also easy to see that differences in individual pinnae could cause quite audible changes in hi-fi sound quality, if such changes were made with tone controls or equalizers. For a sound coming from the side, a slow rise in response starts at 200 Hz, reaches 15 db at 2500 Hz, drops to 1 db at 5 Hz, rises to 12 db at about 7 kHz and then drops to 4 db at 10 kHz. This is quite different from the dead-ahead response and indicates that we are very sensitive to the direction from which sounds originate even if we listen with only one ear. For sounds from directly behind, the pinna causes a dropoff of 23 db between 2500 Hz and 10 kHz. Other radically different frequency responses occur for sounds coming from above or below, and indeed the pinnae seem to be entirely responsible for our sense of center-front sound-source height.

What this means for ambiophonics is that whatever sound we generate must come to the listening position from the proper direction. In theory it would be possible to modify the frequency response of ceiling reflections to mimic side reflections, but such an equalizer would have to be readjusted for each human being. It is much easier to place the ambient loudspeakers around the listener and feed the appropriate signals to them, as described in later chapters. These pinnae effects also explain why launching deliberate or inadvertently recorded rear reverberant hall sounds from the main front loudspeakers in simple stereo systems does not and cannot sound realistic.

Although a one-eared music lover can tell the difference between a live performance and a stereo recording listened to at home, and ambiophonics can help such an individual, it is two-eared listeners that ambiophonics can help the most. Two ears can enhance the listening experience in a concert hall (and life in general) only if there are differences between

AMBIOPHONICS

the sounds reaching each ear, at least most of the time. The only differences the sound at one ear compared to that of the other ear can have are differences in intensity, arrival time and absolute polarity. In an acoustical concert hall or any real physical space, it is not possible for absolute polarity to be inverted at just one ear and certainly not at just one ear at all frequencies simultaneously. Thus we only need to consider what the difference or lack of difference between the ears in sound arrival time and intensity does for listeners, particularly at a concert.

It is clear — since the distance between the ears is relatively small — that at very low frequencies there can be no significant intensity or arrival time differences, regardless of where a low-bass sound originates. At the other, very high frequency extreme, the head is an effective barrier to such sounds coming from the side and, therefore, intensity differences provide the strongest directional cues. At the higher bass frequencies the ear can begin to use arrival time differences to locate a sound. At higher frequencies in the 500 to 1500 Hz region, both time and intensity differences play a role, until as the frequency continues to rise only intensity differences matter. Finally, the ear is sensitive to the arrival time of sharp transients and will readily localize a source based on which ear detects a sudden rarefaction or compression first.

Since two-eared listening is more vibrant than one-eared listening, sound fields that deliver sounds that differ in intensity or arrival time are more exciting, and in concert halls add spatial interest to the event. Thus when we come to consider home-concert-hall design, it is not enough just to maintain the separation of the front left and right channels; it is also necessary to ensure the diversity of all of the signals launched into the home listening space. Correlation is the opposite of diversity, and in the next chapter we will consider the significance of the correlation factors of music and auditoriums.

There is one more relevant psychoacoustic characteristic of the binaural hearing mechanism which does not directly relate to intensity and arrival time. This is the ability of the ear-brain system to focus on one particular sound source out of many. Most of us can, if we wish, pick out just one voice or instrument in a quartet, or, in the classic example, overhear one conversation at a noisy cocktail party. This focusing ability is stronger in live three-dimensional concert situations and weaker when trying to distinguish one voice in a monophonic recording of Gregorian chant at home. The relevance to ambiophonics is that if you can generate a concert-hall sound field real enough to fool the brain, the ability to focus comes into play. At a live concert, distractions such as coughing, subway rumble, and program rattling are much less obtrusive because one can focus on the stage and the music. Likewise at home, such distractions as needle scratch, tape hiss, hum, and domestic noises become easier to ignore if you are immersed in ambiophonic atmosphere. This concentration effect is particularly useful when playing CD transfers of noisy acoustic-era recordings.

CHAPTER 2
CONCERT-HALL SOUND CHARACTERISTICS

In order to recreate a realistic concert-hall or opera-house sound field at home, it is necessary to know how a good music auditorium works. Literally hundreds of papers and books have been written on this subject, and while concert-hall design is now largely based on computer simulation and known acoustic principles, there is still a lot of subjective opinion and art involved.

Concert-hall listeners, not too far back in the auditorium, usually can detect left-to-right angular position of musicians on the stage, can sense depth or the distance they are from the performer, can sense height if, say, a chorus is elevated above the stage, can sense the size of the space they are sitting in, and finally, sense its liveness or, in technical terms, its reverberation time. Some people can also sense where they are in such a space and what is behind them. When listening to recorded music at home, we want our system to provide us with the same sonic clues that the concert hall provides to its patrons.

In this chapter we explore what makes a hall sound good, so that we can determine which features of a hall we must absolutely duplicate at home in order to fool our ears into thinking that we are in a concert-hall space that is palpably real.

AMBIOPHONICS

DIRECT SOUND AND PROSCENIUM REFLECTIONS

First, for a listener in the audience, there must be an unobstructed path for direct sound to travel from the stage to the listener's ears. This direct sound is then followed by early reflections from the back wall of the stage, the side walls of the stage, the ceiling and, to a lesser extent, the floor of the stage. These first or early reflections from the stage proscenium come at the listener from roughly the same quadrasphere as the direct sound, i.e., the front. Depending on the depth, width and height of the stage, and its sound reflectivity, these early proscenium reflections arrive from 10 to 300 milliseconds after the direct sound and are fairly strong.

SOUND - SIGNAL CORRELATION

At this point we must introduce the concept of sound-signal correlation. A piece of music on paper, such as an organ fugue, has a correlation value that represents how the present sound relates to the previously heard sound. The extent of this self-correlation, called autocorrelation, depends only on the composition and instrumental content of the sound itself and the length of time over which correlation is looked for. The autocorrelation value of music played in a real hall will be altered by the amplitude, delay, angle of incidence and number of reflections experienced. Correlation factors go from 0 to 1 where 1 means the next sound is completely predictable and 0 means there is absolutely no relationship between one note or transient and another. We are also very concerned with the correlation between the sounds reaching the right and left ears. This correlation factor is called Interaural Cross-Correlation (IACC). The existence of IACCs less than 1 makes stereophonic and binaural perception possible. Thus, there are autocorrelation factors that describe the signals impinging on a single ear, and there are the interaural cross-correlation factors that describe the sound differences between the ears.

An example of simple autocorrelation properties is the round "Row, Row, Row Your Boat" as sung by two voices outdoors. If we look at the sound over the short time it takes one voice to sing "Row, Row," and the other voice to sing "Merrily" the voices will appear to be entirely uncorrelated. On the other hand, if we look at the relationship over a period of minutes, we would discover a higher value of autocorrelation since each voice eventually sings exactly what the other voice has just sung. If one voice is a tenor and one a soprano, this correlation is weakened, and if the tenor sings out of tune, softly, in French, and is indoors in the next room, the correlation factor begins to approach zero. Most people would prefer to hear such a performance with a correlation factor higher than 0, but still much less than 1, which would imply that the tenor and soprano were singing precisely the same notes and words at the same time, in the same milieu, and in the same vocal range.

AUTOCORRELATION AND MUSICAL SOUNDS

Different types of music have different autocorrelation values when looked at through a window of three seconds or longer. For example, an organ playing in a cathedral will have a significantly larger value than a solo guitar playing outdoors. The reason all this is pertinent to concert-hall sound is that the autocorrelation value of music determines the type of ambient field that will make it sound best. Thus a concert hall may be well designed for orchestral music, but be a horror for a string quartet. The advantage of the home concert hall is that, unlike the real hall, we can, if we wish, adjust ours to suit the autocorrelation value of each musical selection.

SIGNIFICANCE OF HALL IACC

While hall reverberation characteristics are the key factor in coping with autocorrelation problems, it is really the interaural cross-correlation value of the early reflected sounds that largely determines the quality of a concert hall and provides the best aural clues to hall presence. In the concert-hall-ambience world, the IACC value largely represents what happens in the milliseconds after the arrival of a direct sound sample. Hall design research has shown that the IACC should be kept as small as possible (greatest signal difference between the ears for as long as possible) for the most pleasing concert-hall sound. This should come as no surprise to audiophiles who have always believed in maintaining as much left-right signal separation as possible.

To quote Professor Yoichi Ando (Concert Hall Acoustics, Springer Verlag, 1985), "The IACC depends mainly on the directions from which the early reflections arrive at the listener and on their amplitude. IACC measurements show a minimum at a sound source angle of 55 degrees to the median plane." To translate this, the average person's ears and head are so constructed that a sound coming from 55 degrees to the right of the nose, impinging on the right ear, will not produce a very good replica of itself at the left ear due to time delay, frequency distortion and sound attenuation caused by the ear pinna shape and head obstruction. The IACC value for this condition is typically .36, which is a remarkably good separation for such a situation.

Ando points out that 90 degrees is not better because the almost identical paths around the head (front and back) double the leakage and, therefore, do not decrease the IACC effectively, particularly for frequencies higher than 500Hz.

By contrast, if an early reflection or any sound arrives from straight ahead, the IACC = 1 since both ears hear exactly the same sound at the same time, and this is desirable for the direct sound from sources directly in front of the listener.

That is, the direct frontal sounds should be more correlated than any reflective signals that follow in the first 100 milliseconds or so. In general, it is best if no early reflections come from straight ahead. Later on as reflections bounce around in the hall, the IACC of the reverberant field increases substantially. The rate at which this interear-signal similarity increases determines how good a concert hall sounds when a piece of music with a particular autocorrelation value is being performed. That is why a pipe organ sounds better in church than in a disco.

The lesson to be learned from all this correlation stuff is that early reflections in the home listening room should have as much left-right signal separation as the recording allows, that such early front reflections should come from the region around 55 degrees in front of the listener and, unexpectedly, that the normal 60 degree separation of the main left and right loudspeakers is not optimum for central stage located sound sources.

MORE ON EARLY REFLECTIONS

Some front proscenium reflections in the concert hall come from above. However, such vertical reflections are not fully shadowed by the head and strike the pinnae of both ears from pretty much the same angle with the same amplitude and at the same time. Thus these reflections are highly correlated at the ears and, therefore, have little effect in adding to the spatial interest of a concert hall. In our discussions of home concert halls, we will, therefore, assume that early reflections from above are often deleterious, can be safely ignored and, indeed, experiments with raising front reflection speakers overhead show this to be counterproductive.

In general, since music performances tend to take place on a horizontal performance plane, sonic height cues for a listener in the tenth row and farther back are likely to be inaudi-

CONCERT HALL EARLY REFLECTIONS

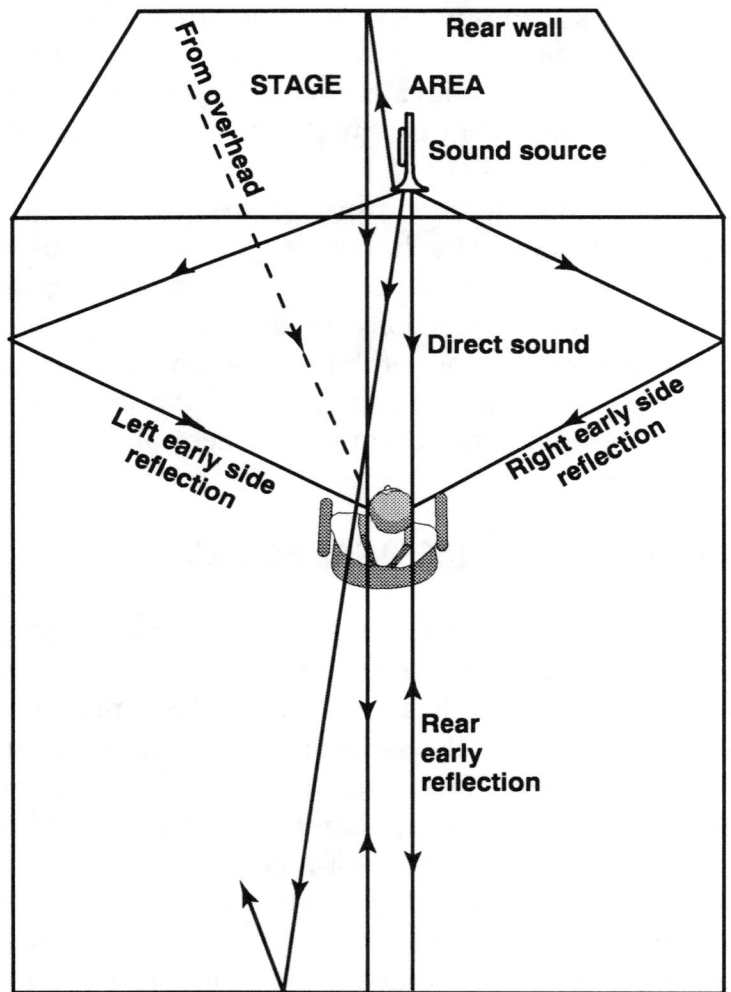

Typical early reflections a concert listener might encounter.

ble. For this reason and because with only two channels there is little that can be done to preserve frontal height cues, we forego height in the ambiophonic concert hall.

To quote Ando again, on early reflections: "The time delay between the first and second early reflection should be 0.8 of the delay between the direct sound and the first reflection." That is, later reflections should be closer together. "If the first reflection is of the same amplitude and frequency response as the direct sound, then the preferred initial time delay is found to be identical to the time delay at which the envelope of the autocorrelation function (coherence of the direct sound) decays to a value of 0.1 of its initial value." Ando found that first reflection delays of from 35 to 130 ms. were preferred, with the exact listener preference directly proportional to the duration of the autocorrelation function or the average time over which the music is related to itself most strongly. That is, listeners prefer later initial reflections for organ music or a Brahms symphony and earlier ones for a Mozart violin sonata. Such a preference is perhaps intuitively obvious: for most organ music, if the first reflection arrived too soon, it would be ineffective, since the same direct note is probably still sounding. We will make use of these rules of thumb when it comes time to set the early-reflection parameters for a given recording in our synthesized concert hall.

To summarize, the front-side early reflections are the most useful in either a real or simulated concert hall and should be centered on 55 degrees. The frequency response of this reflected sound should be similar to the direct sound. If the walls are symmetrical, then the IACC for a centrally located listener is increased, because identical reflections from central sound sources arrive at both ears simultaneously. Our listening room, like a concert hall, can be made more exciting by using an asymmetrical room shape, asymmetrical speaker placement or, best of all, asymmetric early-reflection signal generation. Finally, as many concert hall designers have suggested, strong early reflections from the ceiling and rear walls should be steered or diverted to come from a direction that minimizes the IACC. We will accomplish this by room treatment and by not sending any such undesirable early reflections to the side and rear reverberation loudspeakers.

REVERBERATION

After the mostly frontal early reflections come the rear, ceiling, and rearward side reflections and reflection of these reflections from the proscenium and all the other hall surfaces.

Once these reflections are so close together that the ear, or even measuring instruments, cannot distinguish them, they are called collectively "reverberation," and form a reverberant field. The reverberant field has many parameters that concert-hall designers tinker with and that we will be able to season to taste at home. They are the sound level at the onset of the reverberant field, its density, its frequency response and response changes with time and angles of incidence, its diffuseness (i.e., its directionality versus intensity), its rate of decay and its interaural cross-correlation. Combinations of these reverberant train parameters allow a listener to perceive the liveness and, to some extent, with the early reflections, the volume of the structure. The reverberation preferences of listeners are again dependent on program material. Chamber music and string symphonies usually sound better with shorter reverberation times. (For the record, the official definition of reverberation time is the time it takes for the sound pressure of a single impulse to fall by 60 db or to one-milllionth of its initial strength). Large choral works and organ recitals usually benefit from longer reverberation times, with opera stagings somewhere in between. In numerical terms, reverberation times range from 3 seconds for cathedrals to 1 to 2 seconds for opera houses and concert halls, to .5 to 1 second for bright recital halls. Since the home listener may perhaps have a wide-ranging record collection, we must take care to see that the home concert hall can be quickly optimized for the specific recording being played.

DEPTH PERCEPTION

The ears' ability to detect distance is not as good as that of the eyes'. Depth localization depends on a hazy feeling for absolute loudness, timbre differences with distance (such as high frequency roll-off), time-of-arrival differences between near and far sound sources and, if indoors, the ratio of direct to reflected sound. The first three of these factors are easily captured on recordings by microphones, or can be manipulated by recording engineers, using delay compensation for spot microphones. Nothing in the home concert-hall playback arrangement alters recorded depth perception based on these first three factors.

The fourth depth localization factor is sometimes difficult to preserve. If a recording is made outdoors or with microphones that do not pick up many reflections or much hall reverb, then any ambience added later during reproduction will affect all sound source positions equally. For example, increasing the level of the reverberant field makes the listener feel he is further back in the auditorium rather than increasing the distance between front and rear instruments.

However, as a practical matter, I do not sense any loss of depth perception in my own home concert hall. This may be because, in the average live concert hall, the stage and its shell are so reflective that the direct sound of all instruments, whether located at the front or the back, have about the same ratio of direct-to-reflected sound. This front-to-back stage depth, as opposed to average distance to the stage, particularly for a balcony listener, is not easy to perceive in the typical concert hall. Also, in some recordings, multiple spot microphones are placed so close to their sound sources that almost no differences in the ratio of direct-to-reflected sound of any instrument gets recorded. Ambience pickup is then relegated to other remotely placed microphones, so again all instruments recede together. In the home reproduction system, as in the concert hall, it is unlikely that any lack of differential depth perception will actually disturb the illusion of being there.

CHAPTER 3
STEREOPHONIC SOUND FIELDS

Human hearing using two ears is called binaural and was developed by evolution. Binaural sound is what most of us listen to all the time. Stereophonic sound, by contrast, is simply one man-made method of recreating a remote or recorded sound field in a completely different space. Stereophonic sound fields are almost always auditioned by binaural listeners. The ambiophonics technique was developed to be primarily compatible with stereophonic recordings that consist of two full-range unencoded, discrete channels, one left and one right. One of the basic premises of this book and the research it describes is that the usual two-channel recorded program material contains sufficient information to allow accurate simulation of the binaural concert-hall experience.

That as few as two channels should be adequate can be intuitively understood by simply stating that if we deliver the exact sound required to simulate a live performance at the entrance to each ear canal, then, since we only have two ear canals, we should only need to generate two such sound fields. The questions are why existing stereophonic and binaural recording techniques fall short, and what can be done to make up for these shortcomings.

MONOPHONIC SOUND

Before the advent of stereo recording we had single-channel, or monophonic, recordings. Most recordings were made by using one or more microphones and mixing their outputs together before cutting the record or making a tape. A monophonic recording, if reproduced by two loudspeakers, can be thought of as a special case of stereophonic sound reproduction. It is the case where the interaural cross-correlation factor of the sound is 1 (see next chapter for definition of IACC). In a concert hall, such a signal coming from in front of a listener is sensed directly in front.

Let us now consider a listener in the balcony of a large hall. For this listener, the angle that the stage subtends is very small, the direct sound from the stage is weak because of distance, and the hall reverberation is strong and largely the same at each ear; thus, the players seem to be remote, but still front and center. However, on the plus side, the balcony listener is enveloped in a pleasing but mostly monophonic reverberant field and, therefore, hardly notices that his ability to localize left and right sounds is minimal. The lesson we want to draw from this is that mono recordings can be made to sound every bit as realistic in the home concert hall as stereo recordings, if you don't mind the impression of sitting further back in the auditorium.

The reproduction of monophonic sources via two front loudspeakers is also prone to exactly the same crosstalk effects that result from binaural listening to stereophonic reproduction, but, fortunately, the solution is the same (see below) for both monophonic and stereophonic recordings. Thus it is possible to have realism without separation, but the ambiophonic combination of "you-are-there" ambience with a corrected stereophonic effect is truly ear-boggling.

STEREOPHONIC SOUND REPRODUCTION

In theory a perfect replica of a given concert-hall sound field can always be produced by putting an infinite number of forward-facing microphones at the front of the stage, all the way up to the ceiling. After being stored on a recorder with an infinite number of channels, this recording can then be played back through an infinite number of point-source loudspeakers, each placed exactly as its corresponding microphone was placed. While the performance replication of such a wall would be virtually perfect, the final result would depend on the quality of the room all these speakers radiated into. The major advantages of such an arrangement would be that it works in any listening room (although having a full opera-house stage attached to a small living room might sound peculiar) and almost any number of listeners could hear the full panoply of stage width and depth just as they do in the opera house.

As the number of microphones and speakers is reduced, the quality of the sound field being reproduced suffers. By the time we get down to two channels a great deal has been lost (height cues for a start), but a surprising amount of directional data remains which, if reproduced properly, can be used to effectively reconstitute the original sound field. The most popular two-channel method is the stereophonic technique of reproducing two-channel recordings through two loudspeakers. Stereo takes advantage of one basic psychoacoustic, ear-brain system characteristic, which is that as a recorded sound source moves on the stage from the left to the right, and as the playback signal likewise shifts from the left speaker to the right speaker, the listener hears a sound image move from one speaker position to the other. If identical sounds come from each speaker (the monophonic case above), then a central listener hears a phantom sound that hangs in the air at the halfway point on the line between the loudspeakers. This illusion of frontal separation and space is so pleasing to most listeners that stereophonic recording has remained the standard technique ever since Alan Dower Blumlein applied for his patent at the end of 1931.

THE STEREOPHONIC ILLUSION

However, the illusion created by stereo techniques is far from perfect, even if the highest grade of audiophile-caliber reproducing and recording equipment is used. The first problem is that the image of the stage width is confined to the arc that the listener sees looking from one speaker to another. Occasionally, an out-of-phase sound from the opposite loudspeaker, an accidental room reflection, or a recording anomaly will make an instrument appear to come from beyond the speaker position. These images, however, are almost always ephemeral and often not reproducible. Thus, in non-ambiophonic systems, in order to get a usefully wide stereo effect or stage width with stable left-right localization, the loudspeakers must be placed, in an angular sense, quite far apart, usually forming an equilateral triangle with the listener. As we shall see in the next chapter, it is better if speakers are put closer together so as to mimic the average microphone position and complement the binaural hearing mechanism. With most stereo systems, there is a "sweet spot" at the point of the triangle where the listening is best. This, unfortunately, is what we are faced with when only two channels are available, and the "sweet spot" characteristic is also true of the ambiophonic reproduction technique described below. It is difficult enough to recreate concert-hall sounds from two discrete recorded channels, for one or two listeners in the home, without trying to do it for a whole room full of people.

STEREOPHONIC CROSSTALK

By far the major defect of stereophonic reproduction is caused by the presence of crosstalk from the loudspeakers at the listener's ears. Eliminating this crosstalk widens the stereo soundstage way beyond the position of the loudspeakers, eliminates spurious frequency periodic sound-cancellation effects (comb filter effects) — which interestingly enough are usually undetectable as just changes in timbre — and allows the speakers to be moved much closer together, coming closer to the binaural ideal for angular perception.

In a concert hall, direct sound rays from a centrally located soloist reach each ear simultaneously: one ray per ear. By contrast, for a centrally located recorded sound source, identical rays come from the right and left speakers to the right and left ears, but a second pair of uninvited, only slightly attenuated, longer right and left speaker rays also passes around the nose to the left and right ears.

The problem is that these unwanted rays, which cross in front of the eyes, are delayed by the extra distance they travel across the head. At its greatest, this distance is just under 7 inches. For an average distance of say, 3 1/2 inches, it takes sound one-quarter of a millisecond to do this. A quarter of a millisecond is half the period and, therefore, half the wavelength of a 2000 Hz tone. When two signals, one direct and one a half-wavelength delayed, but of similar amplitude, meet at the ear, cancellation will occur. At 4000 Hz the delay is one full wavelength and the sounds will add. Thus at frequencies from the octave above middle C and up, all sounds add or subtract at the ears to a greater or lesser degree, depending on the original sound source position, the angle to the speakers, the listener's head position, nose size and shape, head size, differing path lengths around the head, pinna shapes and other geometrical considerations. Now note that if the sound source at the recording studio or the listener at home moves a few feet or inches to the left or right, a whole

AMBIOPHONICS

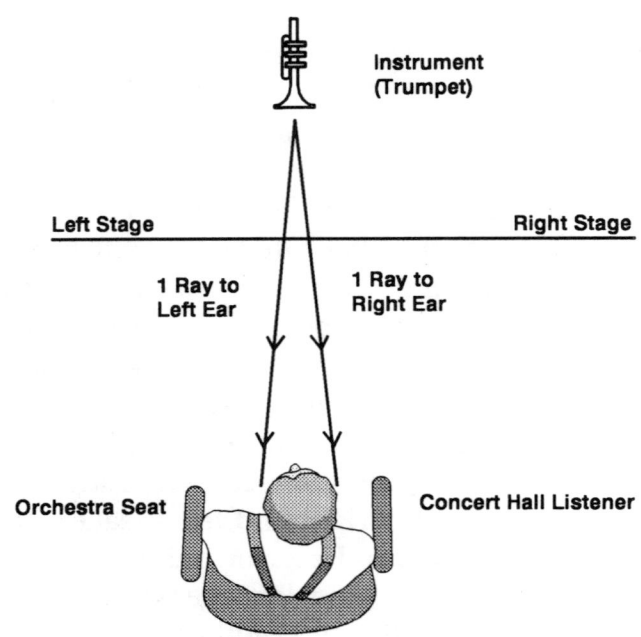

Comparison of live concert hall listening geometry with home stereophonic listening practice showing the additional crosstalk sound rays that cause poor stereophonic imaging effects.

AMBIOPHONICS

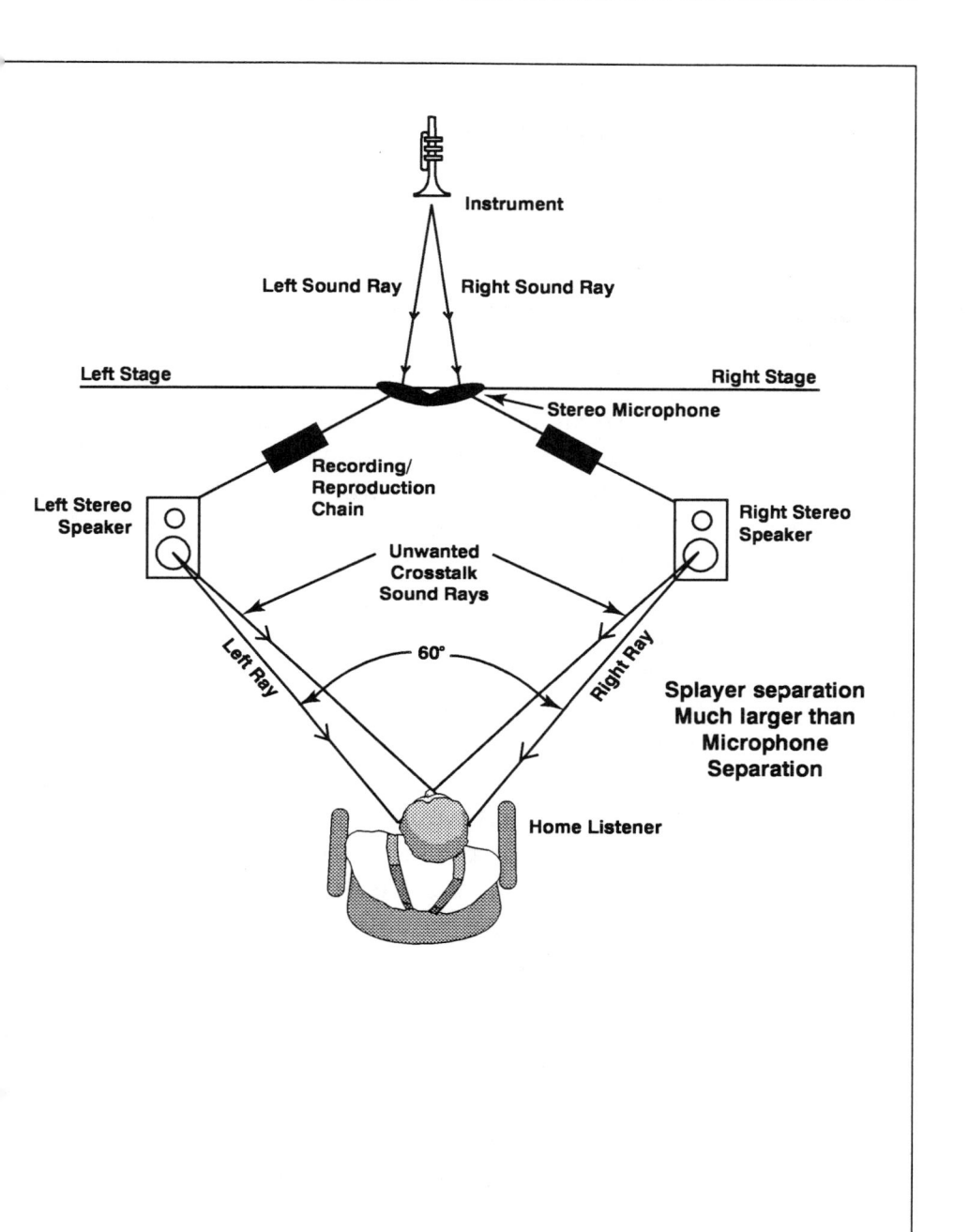

31

new pattern of cancellations and additions at different frequencies will assault the listener. This interference phenomenon is called comb filtering, and largely explains why many critical listeners are so sensitive to small adjustments in listening or speaker position, recording techniques, and relatively minute playback-system electrical and acoustical delay or attenuation characteristics.

Bock and Keele measured comb filter nulls as deep as 15 db for the 60-degree stereo loudspeaker setup. Note that for extreme side images the comb-filter effect is minimal. Thus the frequency response of a normal stereo setup actually depends on the angular position of the original instrument or singer. As indicated above, it is fascinating that these frequency response anomalies are not audible as changes in tone but rather manifest themselves as imprecisions in imaging and a sense that the music is canned. Since the pinnae are quite sensitive to the direction from which sounds come, it is not surprising that the phantom image of a centrally located instrument is often vague or unrealistic. The standard stereo loudspeaker arrangement expects the right ear to interpret one sound from 30 degrees to the right and one sound from 30 degrees to the left exactly as it would a single sound from directly in front. Not likely. This pinna-directional sensitivity effect is another good reason for having the main front speakers as close together as is practical.

Another way of understanding why stereophonic crosstalk is so deleterious, even though the sharpness of the comb-filter nulls makes them inaudible as far as timbre or frequency response is concerned, is again a pinna function. Depending on the location of a sound source and its frequency components, the pinnae produce their own set of comb-filter nulls in the ear canal which, with binaural and broader frequency-response cues, help the brain determine a sound source's location in space. If now a second sound source, at a much different angle but only slightly delayed and highly correlated with the first source, unnaturally impinges on the same pinna, it generates a conflicting set of peaks and nulls, eliminating the pinna as a fully effective image-detecting mechanism.

As an example of how interaural crosstalk and the family of acoustic notches it produces can heighten many listeners' sensitivity to component differences, the case of tube-vs.-transistor amplifiers stands out. There is often a clearly audible difference between a vacuum-tube amplifier and a transistor amplifier used alternately in a given stereo system that is still detectable even after distortion, noise, power and volume characteristics are matched. This difference is usually described in terms of the stereophonic sound stage produced. The image with one amplifier is said to be more transparent, wider, deeper, narrower, shallower, more detailed or less ambient than the other. However, if you listen to just one channel with just one speaker and, even better, one ear, there is of course no stereo effect and these audible differences in sound essentially become impossible to hear and often evaporate entirely, if the amplifiers being compared are of good quality and reasonably matched. You can also do this experiment using both ears listening to one speaker, but that speaker should be a single full-range driver or not have a crossover network in the 250-to-6000 Hz region. This experiment serves to confirm that we are definitely dealing with a two-eared phenomeon and the only logical candidate that presents itself that could produce such an effect is a stereo crosstalk phenomenon—but which one?

The apparent difference in sound-stage imaging due to changing from tube to transistor amplifiers is overwhelmingly due to the different output impedances of these two devices, leading to audible changes in the stereo crosstalk sound field. Vacuum-tube amplifiers have a higher output impedance, sometimes as high as one or two ohms. Thus if two loudspeakers have slightly different reactive midrange impedances, usually due to crossover network design and component tolerances, and the amplifier output impedances are also not precisely matched, often due to tube aging or bias drift, the delay differences in the crossover region between the two stereo channels will be appreciably greater in the high-source impedance vacuum-tube case than in the more constant voltage solid-state case. Note that a phase shift of only a few degrees can shift a stereo crosstalk comb frequency null by hundreds of hertz. Thus the usual crosstalk comb-filter pattern is a function of any asymmetry in the amplifier output impedance, the characteristics and matching of the crossover networks, and any left/right differences in the impedances of the woofer, the midrange speaker and the tweeter. These differences are insignificant when such a speaker is connected to a solid-state amplifier with a very low output impedance, but are exaggerated in the vacuum-tube case, where, say, the left midrange speaker interacts with the right tweeter to produce interaural crosstalk peaks and nulls that are otherwise not present in the solid-state case. Any changes in the interaural crosstalk pattern are interpreted by the brain as a spatial artifact such as more or less stage width, more or less depth, or more or less air.

Vacuum-tube amplifier chaennels differ in their output impedance, depending on component tolerances. These relatively large amplifier-impedance variations interact with similar differential reactances in speaker systems, which also depend on production variables and left-right reactance symmetry, making it almost impossible to predict with any consistency how a given amplifier driving a given speaker will sound in regard to stage imaging. Furthermore, since width, depth, and ambience can only be judged subjectively, both equipment reviewers and audiophiles find it very difficult to consistently detect and define such subtle but audible spatial differences in high-end equipment. To confound audiophiles even more, the owners of wide-range electrostatic or other closely matched speakers without crossovers, or with electronic crossovers, will have a much harder time detecting any of these reported imaging differences between reasonably matched amplifiers. Similar arguments can be made for anything in a system that changes the comb-filtering pattern including, on occasion, vacuum-tube amplifier speaker cables, if length and type are seriously mismatched left to right. Even a small, one-degree phase shift change between the left and right channels at 2000 Hz will cause a shift of 71 hertz in the position of a crosstalk null or peak. This is about one half of a semitone in any octave (assuming a 6 db up-or-down per octave impedance difference). Of course any change in listener position or speaker location causes similar shifts in the crosstalk peaks and nulls and further complicates equipment comparisons by ear.

Both audiophiles and equipment reviewers are likely to find some crosstalk fields pleasing and some less so. Thus there can never be a truly objective resolution as to whether amplifier A is better than B under the usual stereophonic listening conditions. The only way to get consistent results in this component evaluation context or in listening to recorded sound in general is to banish acoustic interaural crosstalk from the listening room, using one of the methods described below. The irregular directional and largely unpredictable frequency response of the standard stereophonic 60-degree listening arrangement would never be accepted in an amplifier, a preamp, a speaker, or a cable. Why such a basic listening system defect continues to be so universally tolerated and studiously ignored is difficult to fathom. Even the relatively less audible absolute polarity problem is taken more seriously.

In addition to the monaural pinna-directional sensitivity, the binaural perception of directional cues depends on both the relative loudness of sound and the relative time of arrival of sound at each ear. Which mechanism predominates depends on the frequency of the sound. Unfortunately, since these delay and stereophonic comb-filter artifacts have an

AMBIOPHONICS

effect extending from below 500 Hz on up, they very seriously impact on both mechanisms and thus impair the ability of the listener to detect angular position with lifelike accuracy. It is also these crossed rays that limit stereo sound imaging to the area between the two front speakers (see below). If we are to achieve anything close to concert-hall realism, we must eliminate these crosstalk effects and provide a correct single ray for each ear. (This is also true for surround-sound systems.) One way to do this is by electronically generating a signal that exactly matches the crosstalk and feeding it in inverted form to the opposite speaker. This technique, called sonic holography by Robert Carver and the Panorama Mode by Lexicon, was also described by Manfred Schroeder and by researchers at Matsushita mathematically in 1981, but it just doesn't work very well in practice. After all, the exact correction required depends on a person's head, pinnae size and shape, the angular placement of the loudspeakers relative to the listener's position, and his/her head position at any given moment. However, even an imperfect correction is quite audible and much better than ignoring the problem. The speaker manufacturer, Polk, has tried to solve this problem acoustically by using a second speaker on each side, offset just enough and inverted in polarity so as to cancel the offending sound from the opposite speaker. Again, personal geometric variables make it too difficult to solve the problem completely in this fashion. Except for Lexicon, at the present writing, none of the these designs have coped with the infinite series problem of cancelling the second crosstalk caused by the first crosstalk correction signal and then cancelling the crosstalk caused by the second crosstalk-correction signal that cancelled the crosstalk caused by the first crosstalk signal, etc. But keep reading—there is an answer that is foolproof.

It is also possible to make recordings that already include crosstalk cancellation signals. This technique has never proved practical because such a recording would have to be played back on a system with a specific loudspeaker position. However, it is possible for recordings to have inadvertent crosstalk cancellation caused by spot microphones or accidental phase inversion or a very strong early reflection. Such recordings can briefly display exceptional stage width. This phenomenon has probably been responsible for falsely convincing many audiophiles that standard stereo is ultimately capable of supporting very wide stage images.

IMAGING BEYOND THE SPEAKER POSITIONS

For sound sources that originate, say, far to the right of the right microphone, we can temporarily ignore the left-channel microphone pickup. In the stereophonic listening setup, the right speaker will send unobstructed sound to the right ear and a somewhat delayed version of the same sound to the left ear. Since these signals are highly correlated and very close together in time, the ear-brain naturally localizes this everyday sound situation to the speaker position itself. Thus no matter how low the left-channel volume is (assuming the electronic crosstalk cancelling method is not being used), the recorded image can never extend beyond the right speaker in standard stereo. If, however, the right-speaker sound ray crossing over to reach the left ear is very much attenuated, then the ear-brain system localizes the sound to the extreme right, well beyond the speaker position and just where the recording microphones said the source was located. Clearly, eliminating the extra sound ray results in wide spectacular imaging. But how does one do this in practice?

USING A CENTRAL SOUND BARRIER

The only present foolproof method of eliminating front speaker comb filter and crosstalk effects is to place a physical sound-absorbing or other opaque sound barrier between the loudspeakers. This barrier should extend to within a foot of

AMBIOPHONICS

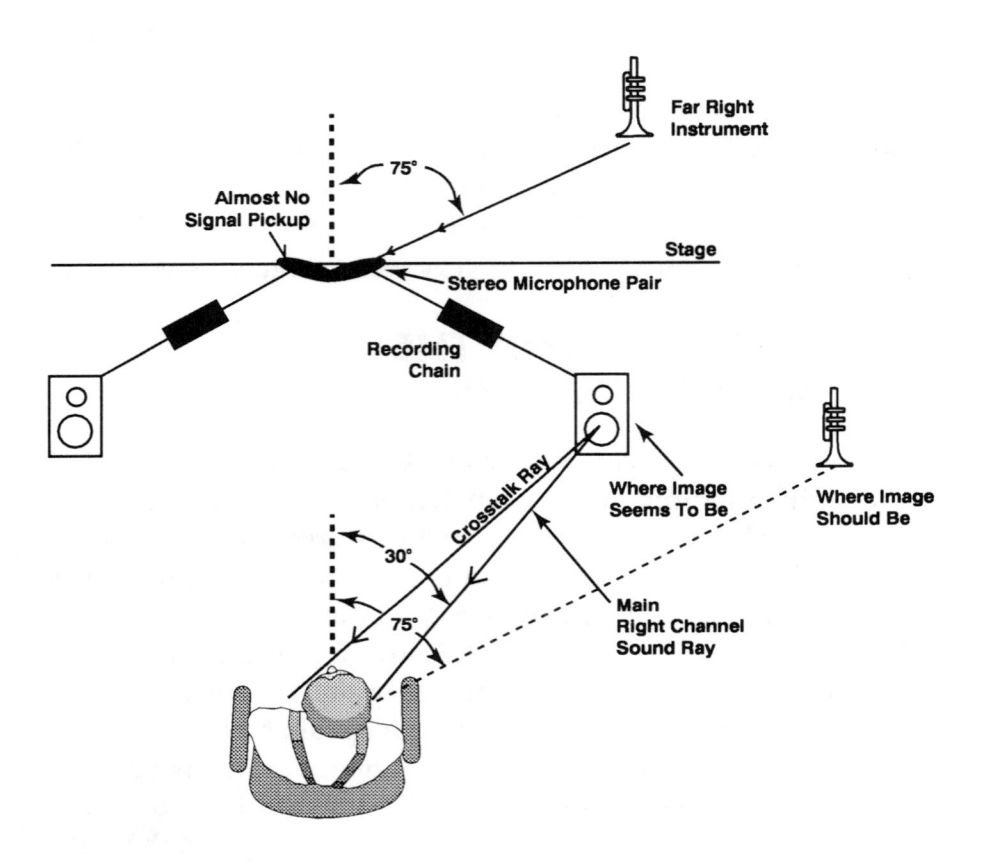

Images in stereophonic systems are restricted to the arc between the speakers because both ears are hearing the same loudspeaker.

the listening position and be cut back at the bottom so that it is possible to sit comfortably at the end of it. The thickness of the barrier is not critical, but should be about six to eight inches, so that when the listener is seated, his or her right eye cannot see the left speaker, and vice versa. The wall or triangle extending back into the space between the speakers is, preferably, made of or padded with sound-absorbing material. This box or panel can be thought of as a collimator for sound. It eliminates all stray rays from the right that might be heading left, and vice versa. Room treatment, as described elsewhere, also helps in this process and is, in turn, helped if the wall is absorptive. Using a triangular sealed box improves absorption at lower frequencies and it can stand on its own.

The use of a reflective barrier to eliminate stereophonic crosstalk was described in 1986 by Timothy Bock and Don Keele, Jr., at that time with Crown International, at the 81st Audio Engineering Society Convention. While ambiophonics prefers the use of an absorptive barrier, their results are still largely pertinent. They determined that a listener can be further back from the end of the barrier if the barrier is wider, the speakers are closer together, and the listener is further from the speaker. Stated as an equation:

$$L = \frac{X(H+T)}{D}$$

where L is the maximum distance a listener's head can be from the barrier, X is the distance from the end of the barrier to the line between the speakers, D is the distance between the centers of the speakers, H is the distance between the ears, and T is the thickness of the barrier. For a worst-case scenario of a six-inch head, a six- inch-thick barrier, an eight- foot barrier length or distance to the speakers, and a speaker separation of three feet, a listener could be as much as 32 inches, almost 3 feet, from the end of the barrier. Thus the barrier does not in any way make listening

AMBIOPHONICS

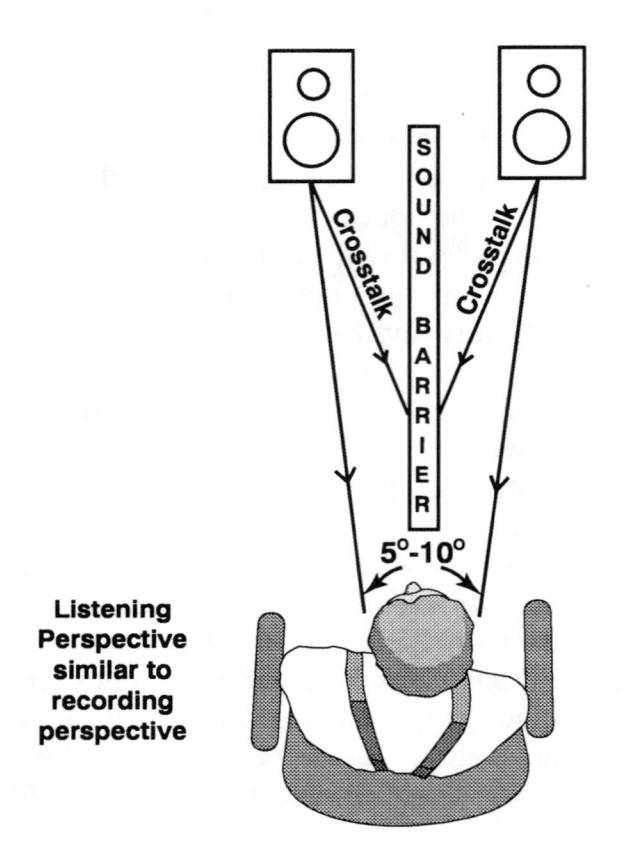

Ambiophonic main front channel listening arrangement eliminating crosstalk and mimicking microphone view.

AMBIOPHONICS

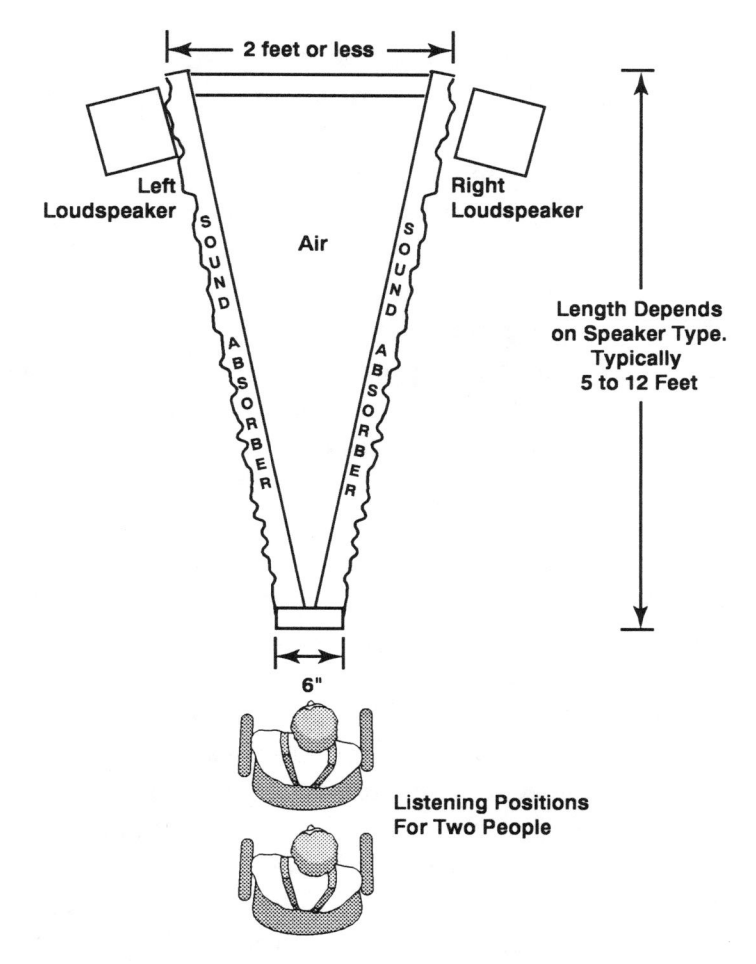

Ideal barrier design to eliminate crosstalk and reduce listening room reflections.

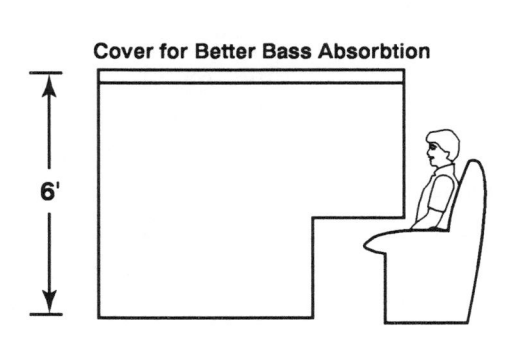

41

uncomfortable or claustrophobic.

Our own personal ambiophonic system geometry allows one to be four feet from the end of the barrier, but at the far end of this range one's head must be more precisely centered. With a four-foot space, two in-line listeners can enjoy the enhanced angular image separation at the same time and indeed the front listener acts as a continuation of barrier for the second listener if the first listener's chair is higher. If in doubt about spacing, the eyeball method is very conservative. As long as no part of the opposite loudspeaker is visible from one eye, excellent separation is guaranteed. Sitting too close to the barrier not only is unpleasant but also results in a loss of high-frequency response if the barrier is as wide as the head, and absorptive.

While we concede that this technique leaves something to be desired where room decor is concerned, the panel can be removed between listening sessions, and at least one manufacturer (MSB Technology, Moss Beach, CA) has come up with a folding absorption panel. It may be that in the future some type of helmet or listening-chair canopy will be available, or even a special transparent earphone that has a built-in very directional microphone that detects only the crosstalk and then cancels it, via the earphones.

While such panels can be bought or custom made, they can easily be fabricated from Sonex or fiberglass panels glued to any surplus office partition or plywood board. Fortunately, neither exact dimensions nor absorption coefficients are critical here.

The improvement such a panel can make is so immediately obvious that if you do nothing else recommended in this book, you should do this. When a barrier is used, the speakers should be moved together so that they face the ears from almost directly ahead. Obviously, directional front-radiating speakers placed close together need less collimation or absorption, and such a location better represents the geometry of an average high-quality microphone or binaural perspective.

STAGE WIDTH

The perception of angular position depends both on the binaural and the monaural abilities of the ears. That is, using only one ear, one can tell from what direction a sound is coming. You can confirm this by covering one ear, closing the eyes, and seeing that you can still detect where sounds come from. The placing of two speakers almost directly in front of the listener, separated by the barrier, results in an image that is at least 120 degrees wide or wider than any normal opera-house or concert-hall stage. But as a sound source moves beyond approximately 60 degrees from midstage, its reproduced image does not continue around. The reason for this seems to be that the monaural directivity sense provided by the pinna convolutions begins to overwhelm the binaural cues. That is, since the sound is actually coming from a speaker in front, it does not have the pinna-altered frequency response of a signal coming from 70 or 80 degrees at the side. It is clear that the pinna effect must begin to dominate at extreme angles, because the change in intensity at, say, the left microphone or ear for a source at the extreme right moving from 70 to 80 degrees is not significant.

In theory it would be possible to change the frequency response, as a function of angular position, using some of the latest surround sound steering technology and assuming an average pinna response or allowing a range of adjustment for individual variations. Alternatively, a model pinna could be placed around the recording microphone. In practice, however, changing stage width is really equivalent to moving one's seat forward or backward a few rows in a concert hall and thus is not of overwhelming significance in ambiophonics.

PINNA SOUND STEERING

One could make a pretty good case for the ear pinnae being more important to our sense of directionality than bin-

AMBIOPHONICS

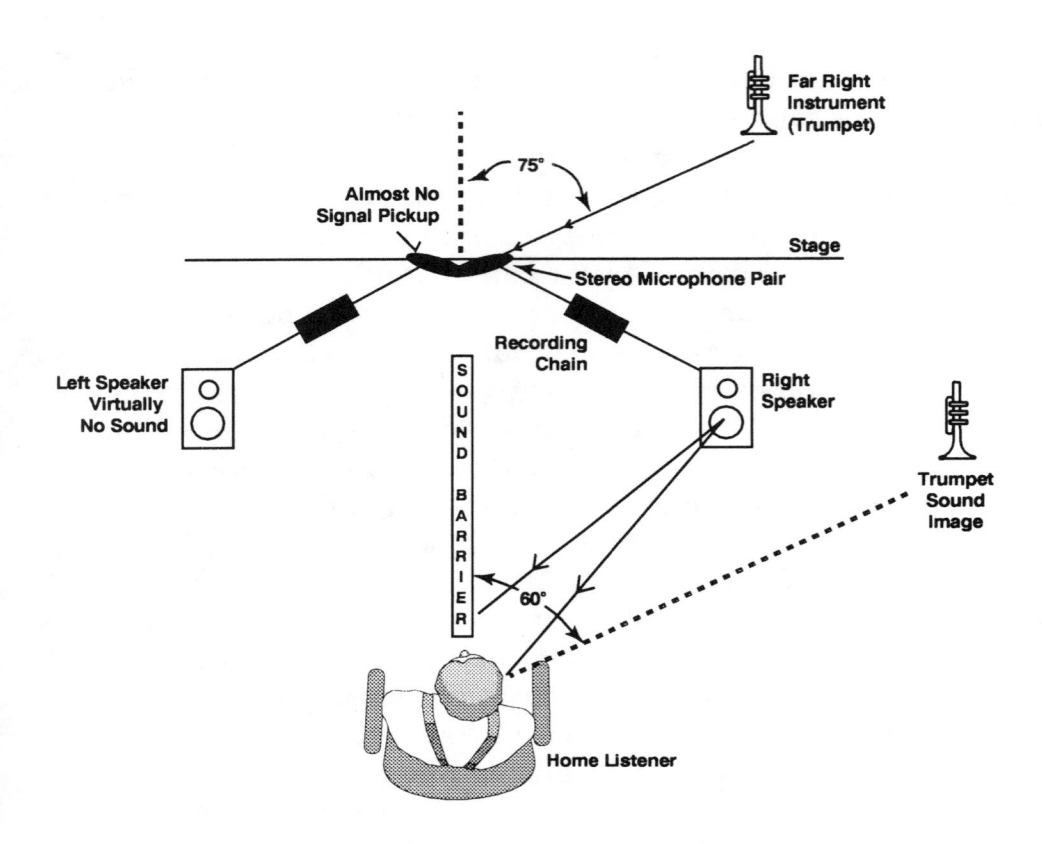

Image in ambiophonic system matches recording perspective because a signal reaching just one ear sounds as though it is coming from the side but pinna effects limit how far to the side the image can be.

aural effects. Arnold Klayman of Sound Retrieval System Labs is a leading exponent of exploiting pinna phenomena in home TV and music systems. The basic idea is that if you equalize a sound well enough, you can make it seem to come from any point in space, regardless of the loudspeaker position. Thus a sound source directly in front can be made to sound as if it is coming from the side simply by frequency-response manipulation. This is done by picking a typical pinna frequency response to frontal sounds, equalizing a front signal with the inverse of this frontal pinna response, and then re-equalizing with the average pinna response to a side-arriving sound. Klayman and others have demonstrated that this basic idea is very effective in both practice and theory, and this type of manipulation is also popular with earphone listening protagonists.

In the context of ambiophonics, however, it runs into several insuperable difficulties. One is that there is no such thing as a single pinna response curve that will apply exactly to every listener. So unless there is some way the equalizer can be quickly tailored to each new listener's pinna shape, the positioning effect will not always be as realistic as it could be. The major musical use for the Klayman SRS technique is to make the ambient sound information on a standard two-channel recording seem to come from the sides or the rear, thus producing a more interesting and realistic sound field. This is accomplished by extracting the difference signal (L-R), equalizing it for an average side or rear pinna, and then outputting the modified reconstituted signal through the main front speakers. This method is fairly effective and certainly better than doing nothing, but it suffers from one major theoretical defect. There simply is no reliable way to differentiate recorded left and right reverberant information from unencoded (or, for that matter, encoded) main front left and right sounds. For instance, an instrument on the extreme left, recorded in a dry studio, will generate a very large L-R signal, even in the absence of ambience, which would then be equalized improperly to sound as if it were rear or side ambience. Also, such L-R or R-L pinna-equalized ambience must of necessity be highly correlated (or usually monophonic) and therefore less realistic than it could be using the ambiophonic method. One could use the logic steering circuits of the surround-sound world or, as Klayman has done, provide matrix volume controls that can be adjusted manually for each recording to correct for these errors. However, if little ambience or reverb is on the record, the reproduction remains lifeless, and other problems arise when monophonic recordings are played.

The ambiophonic method, in contrast, accomodates to the pinna effect by launching the ambient signals from more or less the same directions as the sound would arrive at the ears in a typical concert hall. Thus there is no need to customize pinna equalizers to a particular ear, and in ambiophonics there is no need to extract ambience from the recording itself or even require that the recording have ambience. However, let us imagine that we have a six-track recording, with two dry main front channels, two side ambience channels, and two rear reverberation channels, but we have only two front loudspeakers. One could then pinna-equalize the side and rear signals, mix them with the front signals, and play the combination through the two front speakers. Voilá! If a crosstalk barrier is in place between the front speakers, one can have a pretty good ambiophonic system using just two speakers. This can be tried using either the JVC 1010 or two Lexicon CPs to simulate the ambience channels, which can then be processed by two Klayman devices and mixed to one front speaker, or kept separate and sent to three pairs of front speakers. Except for experiments, this hybrid is too cumbersome to use routinely, but with the new multi-channel recording formats such as AC-3, DTS, etc., such speaker-conserving methods may become quite useful, even if not ideally perfect, particularly in cars, bedrooms, bathrooms, patios, and kitchens.

BINAURAL RECORDING REPRODUCTION

It is logical to suppose that if a recording is made using a dummy head complete with ear pinnae, and microphones at the opening of the dummy's ear canals, and that if you then listen to such a recording using earphones fitted into the ear canal, a perfect "you-are-there" sensation will be produced. Surprisingly, this is not the case. The sound image seems to exist inside the head stretching from ear to ear and nose to rear skull. While no one seems to have determined definitively why this is so, it is clear that each human being has his own shape of pinna and head size, and since no individual's head matches the dummy head closely enough, perhaps the brain decides the sound field is not real.

Now let us make a recording where forward-only sensing microphones are placed on a dummy head slightly larger than most human heads, but without dummy pinnae. Now play this recording back through front loudspeakers separated by the wall that eliminates crosstalk and, the image is no longer in the head. Nothing is perfect, however, and the "surround," or ambient part of the binaural experience, is impaired. For music, though, the ambiophonic method of creating simulated reflections and reverberation produces a magical illusion when combined with the razor-sharp front-horizontal imaging of binaural recordings.

LOUDSPEAKER OUT-OF-PHASE EFFECTS

In virtually all stereo systems it is necessary for the right and left main speakers to be in phase. Phase in this case means that if identical electrical signals are presented to each speaker, the speakers will both generate a rarefaction, or both generate a compression in response to a simultaneous and identical input pulse. When a monophonic recording is played through a pair of out-of-phase loudspeakers, the sound at the ears lacks bass, the phantom center image is not present, and a hazy, undefined sound field seems to extend far beyond the speakers to the extreme sides and sometimes even rearwards. Similar effects are also audible using stereophonic sources.

These subjective effects can be better comprehended now that we understand all about stereo crosstalk. It is clear that equal but out-of-phase very-low-frequency signals with wavelengths much longer than the head will always be out of phase at either ear and, therefore, will always largely cancel. This factor accounts for the thinness of the sound.

At somewhat higher bass frequencies the cancellation is not total, but the left ear hears pure left signal from the left speaker; that is reduced somewhat by the now only partially out-of-phase crosstalk from the right speaker. Similarly, at that same instant the right ear is hearing a reduced but pure right-speaker sound that is similar but not identical to the pure left-ear sound because they are largely out of phase. We know that a sound heard only in the right ear seems to come from the extreme right and a sound heard only in the left ear seems to come from the extreme left. This phenomenon is still operative even if the two sounds that come from the sides are identical in amplitude and timbre. Thus, one can easily hear two identical bells as separate left and right sound sources. If, however, we exchange the bells for pure tone oscillators, then we can hear the oscillators only as separate sources when they are not precisely in step. Since our signals are out of phase they are not identical in time.

Thus the inadvertent crosstalk elimination that occurs at mid-frequencies widens the perceived sound field from out-of-phase speakers. As the frequency increases, instead of cancelling, the comb-filtering effect predominates and the position of the images becomes frequency- and therefore program-dependent, changing so rapidly that no listener can sort out this hodgepodge of constantly shifting, very wide images, and most listeners describe this effect as wide but diffuse or unfocused.

The use of the barrier does not eliminate the need to keep all speakers in the room properly phased. The barrier is not effective at very low bass frequencies and so the bass thinness effect, while less apparent, remains.

However, the audibility of the out-of-phase effect using a barrier is much reduced. The out-of-phase stage image now extends from the speakers outward. That is, sound sources at the extreme right and left create an image as they did before, when the speakers were in phase. This makes sense, since we are, as before, listening to one sound source with one ear. For monophonic or central sound sources, however, each ear is hearing a signal that is distinctive because the signals are out of phase and, therefore, the ear localizes each sound as originating from their respective speakers. The phantom center image does not form and the infamous hole-in-the-middle appears. In the ambiophonic arrangement, however, the front speakers are very close together. Therefore, the middle hole is almost nonexistent and the bottom line is that, except for extreme bass response, front speaker phasing in the ambiophonic system is not as critical as in stereo, but should still, for best results, be correct.

ABSOLUTE POLARITY

When an instrument produces a sound, the sound consists of a series of alternating rarefactions and compressions of air. The sonic signatures of such acoustic musical instruments are determined by the pressure and spacing of these rarefactions and compressions. Electronic recording and reproduction have now made it possible to turn rarefactions into compressions and vice-versa.

The significance of this to the problem of establishing a home concert hall is not entirely clear. But many people can hear a difference between correct and incorrect polarity.

Therefore, care should be taken that all amplifiers, speakers and ambience synthesizers taken together do not, in the end, invert. Since acoustic reflectors do not invert polarity, the key early reflections, at least, should be delivered to the ears with the same polarity as the direct sound which is, one hopes, also of the correct absolute polarity.

If you cannot tell one polarity from the other in your own domestic concert hall, don't despair. For most people, polarity is audible only when special test signals are used. One possible reason for difficulty in this regard is the nature of many instruments. A listener to the left of a violinist hears one polarity, while a listener to the right hears the other polarity (assuming the string is vibrating in the same plane as the ears of both listeners). But no matter where you stand around a trumpet you get the same polarity. The inverted polarity sound in this case is inside the trumpet. Indeed it has been reported that test subjects are more likely to hear polarity differences where wind instruments are involved.

On balance, we would have to say that it does not pay to agonize over the absolute polarity effect unless you are certain that you or your friends are sensitive to it.

CHAPTER 4
CHOOSING AND PLACING LOUDSPEAKERS AMBIOPHONICALLY

We wish now to apply the rules of good concert-hall design to the choice of home concert-hall loudspeaker characteristics and speaker placement. Let us assume that we have available good quality software-generated left and right early-reflection signals, side-reverberation signals, and rear-reverberation signals. Let us also assume that our listening room is treated well enough to eliminate room reflections at the listening position and that the crosstalk-preventing sound-absorbing panel is in place in front of the listener.

There is one general characteristic that applies to all the loudspeakers used in a simulated concert hall: all speakers should be as directional or focused as possible, so as to reduce the number and the level of stray listening room reflections. Remember, no practical room treatment is ever fully absorptive, so the less there is to absorb in the first place, the better the sound.

THE FRONT SPEAKERS

The front speakers should be placed almost directly in front of the listener with each speaker aimed at its respective ear. Front main speakers should be as directional as possible. In theory, the ideal speaker for this purpose would behave like a flashlight, with a sound beam emanating from a single point at ear level and the rest of the room in deep shadow. The more focused the main loudspeaker is, the less sound absorption treatment the room requires and the more effective the crosstalk barrier is.

The angular separation of the speakers as seen by the listener should correspond to the recording microphone perspective as closely as possible. Unfortunately, almost every recording is made with a different microphone spacing or even with many microphones, so this is difficult to accomplish precisely. Most audiophile recordings, however, are made using dummy heads or very closely spaced microphones, and, in our experience, an angle to the speakers of 5 to 10 degrees on each side of the listener works well, on average. Although not relevant in a simulation context via microphones, in a concert hall almost all the direct sound reaches the ear from a direction seldom more than plus or minus 25 degrees, or even less if sitting back in the auditorium. (One day we hope to install a track and cart, so that the front speakers can be easily and quickly moved in and out to match different microphone setups. But, on second thought, we leave this experiment to you, the reader, to carry out.) Otherwise, except that the front speakers used should be capable of reaching concert-hall volume, the normal speaker-selection criteria of good frequency response, low distortion, reasonable time coherence and affordable price apply.

FRONT EARLY-REFLECTION LOUDSPEAKERS

The front early-reflection speaker pair should be placed at the critical plus or minus 55 degree angle to the listening position where the ear is most sensitive to such spatial cues. The ideal speaker here is one that radiates to the listening position from as large an area as possible. Tall electrostatic or ribbon loudspeakers are excellent in this appli-

AMBIOPHONICS

Glasgal Domestic Concert Hall

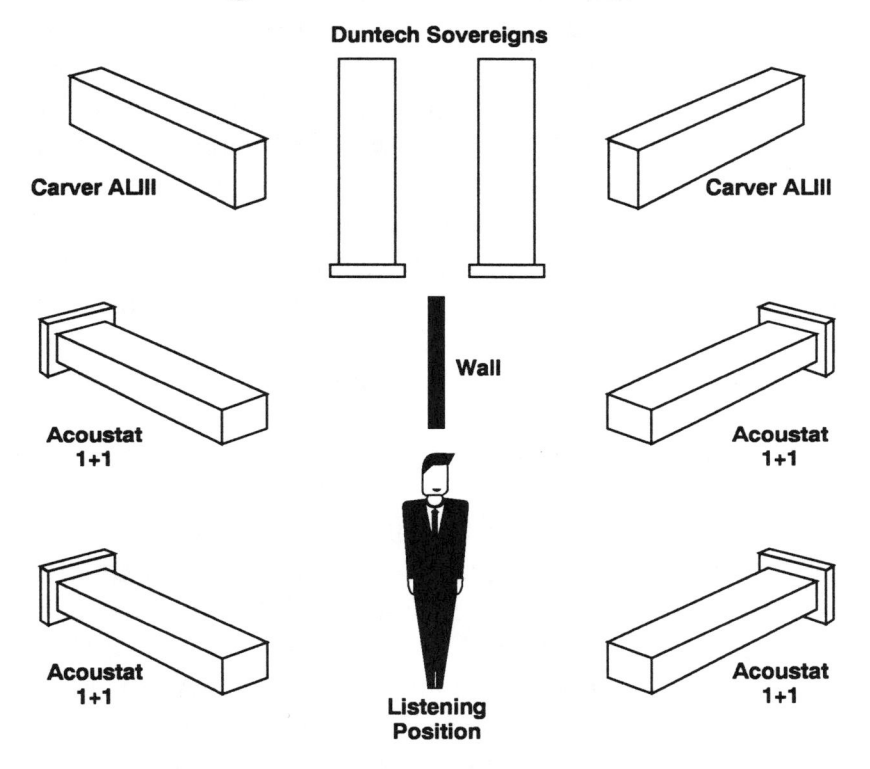

This particular arrangement of tall speakers has produced an exceedingly realistic ambiophonic "you are there" sound field.

cation as they are very directional, but since they are dipoles and also radiate rearward, they must be used with plenty of sound-absorbing material behind them. A useful property of such large-area full-range sound radiators is that they provide significant diffusion without invoking spurious room reflections, as physical diffusion panels would. By leaning tall speakers at a 45-degree angle off the vertical, the ambient signals arrive at the listening position from plus or minus five degrees in both the vertical and horizontal planes. This corresponds to the situation in a real concert hall, where the predominant early reflections arrive from slightly different side directions (but still hopefully around 55 degrees) because the originating sound sources are spread out on the stage and have various angles of incidence and, therefore, reflection. In the home environment, the computer-generated early reflections are the same for all the right-signal and for all the left-signal instruments. This spreading of the apparent reflection image would seem to detract from the concert-hall ideal. But just as the perfect Philharmonic Hall has yet to be built, our home room may be real but not the best. We can only say that for us, leaning the speakers is better. As discussed below, the reverberant field needs to be as diffuse as possible and, therefore, to the extent that either recorded reverberation or synthesized reverb is present at these front-side loudspeakers, there is and additional benefit to being long and tilted, providing both vertical and horizontal dispersion without risking significant room reflection or diffusion of the main front signals.

SIDE AND REAR REVERBERATION LOUDSPEAKERS

The side and rear speaker pairs preferably are fed with four largely uncorrelated synthesized reverberation signals. Since in a concert hall reverb reaches the listener from virtually all directions, the ideal speaker would be a set of thin squares which could be hung on the wall and glued to the ceiling, all aimed at the listening position. I find, however, as above, that large electrostatics or ribbon speakers do an excellent job, particularly if leaned at a 45-degree angle to the floor so as to provide both horizontal and vertical dispersion. Do this tilting in an asymmetrical fashion to reduce incidental left-right correlation. One could also use multiple small, inexpensive box speakers arranged on pedestals in a random roller-coaster pattern around the rear half of the room. Again, in theory, each reverberation sound source should have its own independent reverberation generator, but in practice four speakers (left-side, right-side, left-rear and right-rear) seem to be enough to fool the ear-brain system. If you do use tall speakers and tilt them, the listening room arrangement visually resembles a tipsy Stonehenge.

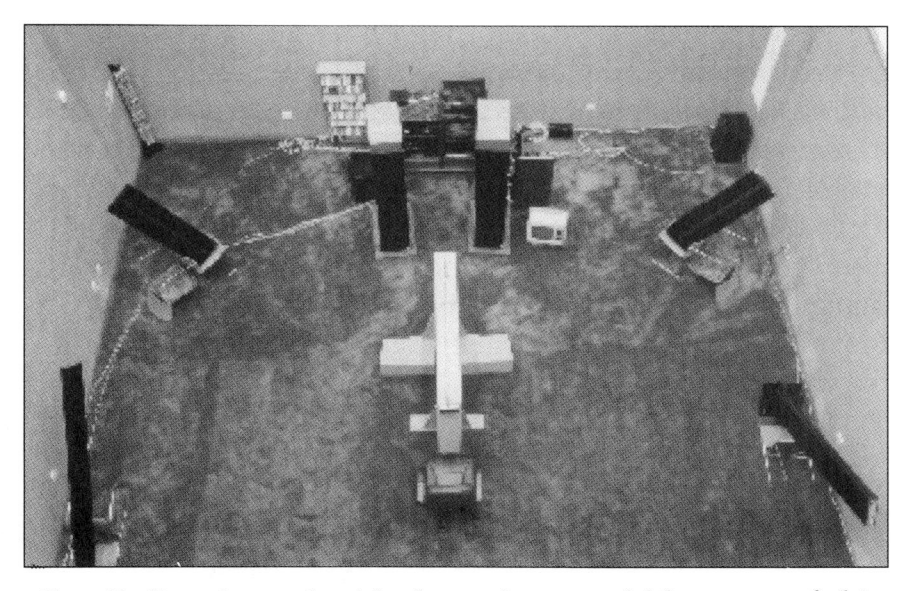

Since the power handling and frequency-response requirements of the reverberation speakers are comparatively modest, this is a good area in which to cut costs. Also, there is no reason why the speakers need to match if they can still be reasonably balanced.

Since the rear reverberant field often has a strong vertical component coming from the auditorium balconies and ceiling, we have found it advantageous to add a second pair of rear speakers elevated as much as possible and driven by their own amplifiers. Ideally these elevated rear speakers would be driven by an independent rear ceiling reverberation synthesizer to provide a richer simulation and a better match to concert-hall design theory, but "better real" is not more real than "real" and this suggestion is, perhaps, gilding the lily.

CHAPTER 5
TUNING THE LISTENING ROOM FOR AMBIOPHONICS

Turning a family room, spare bedroom or rec room into an acoustically viable environment for a genuine home ambiophonic experience need not require a big budget, a building permit or even a single carpenter. The trick is to understand what factors degrade sonic realism in non-purpose-built audio rooms, and then to do something about them.

The possible causes of acoustic disappointment are many, but, encouragingly, experience shows that most home media rooms suffer from insufficient absorption and poor placement of speakers, equipment and furniture.

THE EVIL THAT ROOMS DO

While most speakers can be aimed toward or away from the listener, all loudspeakers spread their output to some extent, like flood-lights illuminating both subjects and surroundings. A speaker firing directly at the listener will also direct sound sideways, up and down, even backwards. In a typical untreated room, this "unaimed" energy hits a wall or cabinet and bounces back toward the listener a split second after the direct launch. (Think of these delayed versions as the acoustical cousins of multi-path "ghosts" on a TV screen.) Presented with a succession of time-delayed, tonally altered and spatially scrambled versions of the direct sound, the brain has an insuperable problem to solve. As a first resort, placing absorptive material — say, 2- to 4-inch thick batts of 3-pound-per-cubic-foot Fiberglass — on the side walls flanking the front speakers tends to restore sonic clarity by reducing the number and amplitude of these early reflections.

As unwelcome as front side-wall reflections are, they are not nearly as nasty as reflections ricocheting up from the floor directly in front of the listener, or down from the ceiling above. That is because the earliest and strongest floor and ceiling reflections come from a middle region between the speakers, in effect beaming monaural information directly at the listener. The audible effect of this strong "anti-stereo" component is to blur imaging and strip away whatever sense of spatial realism the recording contains. A Fiberglass absorber on the ceiling between the speakers and the listener will usually soak up the offending ceiling reflections; a carpet and thick pad will help tame the floor bounce.

THE FURTHER PERILS OF UNTREATMENT

The average untreated living room has a reverberation time of about six-tenths of a second. Since a recital hall could have a reverberation time of as little as eight-tenths of a second, and even concert halls can be in the 1.5 second range, the typical home listening room reverberation time is surprisingly significant compared to the halls in which music is performed. Let us assume that we are playing a recording of a large choral work that has a normal ratio of direct sound to hall reverberant pickup. When such a recording is played in a typically small, live, home environment, the direct sound stimulates the room to produce a reverberant field that tells the brain that the room is small and bright. But then the recorded hall reverberation reaches the ears and tells the

AMBIOPHONICS

brain that the room is large and dry. When you add to this the pinna effects and the extra discrete early room reflections that further confuse the brain, it is no wonder that recordings never seem to sound realistic no matter how much we tweak our systems.

Unfortunately the behavior of home listening spaces is not swayed by the cost or cachet of your cables, amplifiers, and loud-speakers. The plain fact is this: The speakers and the room in which they sit form an acoustic system, and in an untreated room, the latter contributes the lion's share of what you hear. Speaker drivers generate sound waves—changes in pressure—by their rapid in-and-out motion into two different enclosures: the speaker box and the room. Engineers draw on an arsenal of high-tech tools—computers, lasers, accelerometers—to help them zero in on the effects that the size, shape, bracing, padding, and other details of the enclosure have on the anechoic speaker response.

The other enclosure, the listening room, is far more critical to sound reproduction, both stereophonic and ambiophonic; after all, it's larger and our ears are located in it—yet it is rarely the beneficiary of anything approaching the same level of expertise, technical firepower, or plain old-fashioned care. Its acoustical behavior is unknown, uncontrolled, and highly unlikely to replicate the sonic richness—the colors, textures, shadows and shapes that the recording engineers, producers, and artists sweated over in the studio.

The typical residential room:

- makes a droning roller-coaster out of the system's bass response;

- corrupts the perceived tonal quality of instruments and voices;

- scrambles stereo imaging (the perceived location of sounds);

- imposes its own reverberant sound field and treats some frequencies differently than others;

- adds an annoying buzzy or rattle-like quality to percussive sounds;

- creates unpredictable acoustic hot and cold spots; and

- buries the low-level nuances that give music life and believability in ambient noise grunge.

Most people do not have the mechanical building skills to construct or remodel a room to make it suitable for ambiophonic listening. Some can afford to hire an acoustical contractor to handle the design and all the work. For those who can do home improvement projects themselves, the rest of this chapter can serve as a recipe.

The idea is to do as much as you can afford or have the patience to do. Even a partial taming of the spew is beneficial. We know from direct experience that putting four-inch Armstrong fiberglass panels on four walls and all doors, a thick rug on the floor, removing all unnecessary furniture, and buying a central barrier panel does the job quite well at minimal cost. Acoustic tile on the ceiling is advisable if the ceiling is very low.

OVERCOMING THE CENTRAL BARRIER PREJUDICE

A major impediment to implementing the ambiophonic method is the reluctance of serious home listeners to do any room treatment, or particularly to use a central barrier wall between the main front speakers extending to within a foot or so of the listening position. For the overwhelming majority of even audiophile listeners this interior decorating problem seems insuperable. Fortunately, new materials in designer packages ameliorate some of these aesthetic problems. We

AMBIOPHONICS

would first make the observation that when stereo appeared, similar decorating objections were made because placing a second speaker, running a second wire to it, and having to move a chair to a position between the speakers seemed incompatible with living-room decor. Now we have six or even eight loudspeakers in a surround-sound living room, plus in some cases, a video projector in the middle of the floor or dangling from the ceiling, up to a ten-foot screen going up and down, and, in many audiophile listening rooms, room tunes, sonex panels, an assortment of diffusion devices, a large subwoofer cabinet, exotic looking speaker stands, and structural steel equipment cabinets.

Well, as far as adding the central barrier goes, the news is good. Three companies are now making absorption panels that not only do the job at a modest cost, but in all three cases the panels are easily removable once listening is over. RPG Diffusor Systems of Upper Marlboro, Maryland, makes a device called a Pillobaffle. A single pillobaffle is a chubby rectangle 6 feet by 2 feet by 3 inches thick. A single pillo weighs only six pounds. If hooks are placed in the ceiling a pillo can be hung from a short chain whenever ambiophonic listening is desired. A second pillo can be hooked to the first to produce a wall 6 feet by 4 feet, which is more than adequate. Each pillo costs less than $100 and comes in a choice of yellow, green, blue, black, gray, or cream, or you can use a pillo case of your own design.

Pillobaffles can also be wall mounted and so could be used to tame the worst of the specific room reflections. According to RPG, a single wall mounted baffle can absorb 10 Sabines (see below) at 125 Hz (assuming a 12" air space between the baffle and the wall surface), and over 22 Sabins, hanging from the ceiling, in the critical image frequency band from 500 to 4000 Hz. To make the wall six inches thick, two more pillos can be attached to the first two, increasing the absorption and the allowable distance from the edge of the barrier to the listening position. Such pillobaffles, or

AMBIOPHONICS

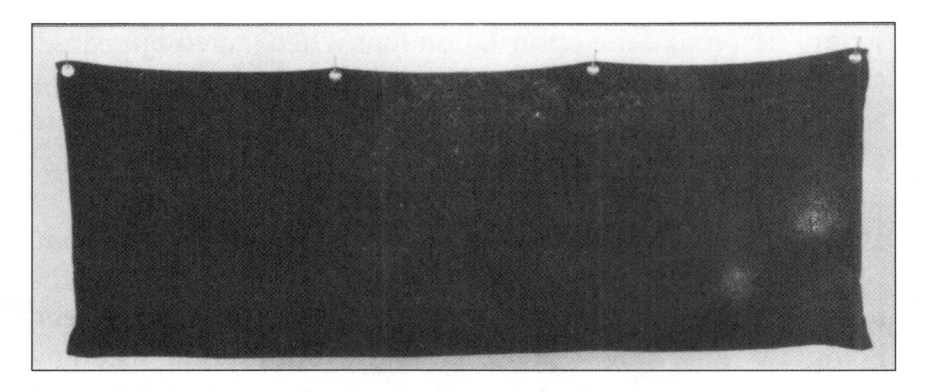

indeed other panels, could be motorized as the new TV screens are, and come down from the ceiling, up from the floor, or out of a front wall.

The Sabine is the unit of sound absorption and it is computed by multiplying the area of an absorbing surface in square feet by its absorption coefficient. The absorption coefficient is simply the fraction of sound that is absorbed by the material at a particular frequency or over a band of frequencies. Thus a window open to the outside swallows up any inside sound that passes through it and, therefore, has the highest possible absorption coefficient of one. If the window is one foot square, its total sound absorption is one Sabine. Ten square feet of 4-inch thick glass fiber could absorb some 9.5 Sabines at 500 Hz and higher, but only about 7 Sabines at 100 Hz. A 660-cubic-foot room (10x14x19) would need approximately 700 Sabines of absorption to get down to a reverberation time of .2 seconds. Using 4-inch fiber wall panels, the area requiring padding would be in excess of 700 square feet, or about half the surface of the room allowing for the small absorption contributed by untreated surfaces, rugs, and furniture.

Acoustic Sciences Corp. of Eugene, Oregon, makes a handsome free-standing acoustical panel called the NA-4 Shadow Caster. This panel weighs 53 pounds, is 5 feet high, 2 and a half feet wide, and 4 inches thick. They are normally sold in pairs for less than $700, which is just as well, since

AMBIOPHONICS

one side is absorptive and the other reflective. By putting the panels back to back with their reflective sides touching, a high performance absorption barrier is created with sufficient width to permit the listening position to be more than 2 feet away from the edge of the panel pair, thus allowing ample foot room. One could also assemble such a wall using the same company's individual tube traps, which would then be easier to move out of the way and store between listening sessions. A single tube trap provides about 15 Sabines and one NA-4, 60 Sabines.

For the ultimate in removability, consider the folding Acoustic Screen from MSB Technology Corp., Moss Beach, California This panel is 6 feet tall with three 2-foot sections. Its maple frame is covered with an acoustic grill cloth that conceals three separate internal damping layers. One screen costs $400 and only one should be necessary. While not quite as absorptive as the RPG or ASC devices, in this application even a much less than perfect central barrier produces a dramatic effect. Additional folding panels could be used to eliminate room reflections if the walls of the room cannot be treated in the normal way.

The central barrier provides benefits that are not directly related to imaging. The absorptive barrier acts to attenuate any stray reflections bouncing around the front part of the room, and it is even large enough to provide damping at relatively low frequencies. In general, the use of a barrier noticeably reduces the reverberation time of the listening room and even makes internal room noises from equipment, fans, etc., less obtrusive.

THE AMBIOPHONIC ENVIRONMENT

The ambiophonic experience depends, more or less equally, on the positive contribution of an ambiophonic playback system, and the lack of contribution by the playback environment itself. Put another way, playback acoustics and "ambiophonicity" are related inversely: the less the playback environment imposes its own personality on the aural mix, the more genuinely ambiophonic the experience can be. This less-is-more phenomenon is the result of the ear-brain system not having to labor at resolving two conflicting sets of acoustic cues: the concert hall (as presented by the play- back system) on the one hand, and the local playback environment on the other. The less adulterated the set of cues, the more persuasive the experience.

At the most basic level, the requirements for an ambiophonics-friendly listening room are quite straightforward:

- low background noise;

AMBIOPHONICS

- high absorptivity, leading to broad-band room reverberation times below .2 seconds: and

- lack of acoustical anomalies (e.g., modal degeneracies) at the listening position.

BACKGROUND NOISE

Wallace Clement Sabine, the father of architectural acoustics, noted nearly a century ago that halls exhibit the same basic sonic behavior at very low sound levels as at very high. If you are an active concertgoer, you may have noticed that concert halls show their distinctive sonic personalities even during those hushed moments when the maestro mounts the podium and raises his baton. There is no confusing the silence of Chicago's Orchestra Hall for Milan's Teatro della Scala or Vienna's Grosser Musikvereinsaal.

Recreating in a residential setting the characteristic sound of a real hall begins with getting that "silence" right. Unfortunately, the typical home is neither designed nor constructed to allow the ambiophile to hear the desirable level of sonic detail. If you turn off your play-back system, shut the windows and doors, and just listen to your listening room for a few minutes with eyes closed, you'll be surprised at how much noise is really there. Its an endlessly rich, constantly changing sonic backdrop of cars passing, kids playing, appliances whirring, lights buzzing, plumbing whooshing and heaters or air conditioners humming. As interesting as all this sonic activity can be, it conspires to smudge the rich, silent spaces between the notes on your recordings, robbing you of the nuances that would otherwise contribute to the near palpability of the ambiophonic or stereophonic experience. Acousticians have developed a sort of numerical shorthand to describe background noise levels. Known as "noise criteria" (NC) curves, and usually specified in increments of 5, from NC-70 (extremely noisy) down to NC-15 (very quiet), these curves are weighted to account for the fact that the ear is less sensitive to low frequencies than to high. The curves' numerical designations are arrived at by taking the arithmetic average of sound pressure levels at 1 kHz, 2 kHz and 4 kHz. A useful target for a purpose-built ambiophonic listening room is NC-20, a spec often encountered in the design of professional recording studios. Note that, should NC-20 prove beyond reach, NC-25 and even NC-30 can still yield acceptable performance, with NC-35 a minimum standard for a legitimate ambiophonic experience. Subjectively, moving up 10 points on the NC scale marks roughly a subjective doubling of noise, e.g., NC-35 sounds roughly twice as noisy as NC-25.

It is difficult, if not impossible, to prescribe a course of action leading to a given NC specification without first knowing what the existing conditions are. Achieving NC-20 performance in a suburban or rural residential setting may prove possible, while even NC-35 may prove impossible in noisy urban settings. Usually heating, ventilating and air conditioning (HVAC) systems are the most notorious noise sources in residential listening rooms. One inexpensive solution here is to temporarily turn off such HVAC systems during serious listening sessions.

RESTORING THE PEACE

To keep noise out of your ambiophonic listening room (and to keep the room from becoming a noise source), you might think to stuff as much dense, sound-absorbing material into the cavities of your wood-stud and gypsum board walls as they'll hold, but the result of your efforts will be a barely measurable increase in the commodity we're looking for: transmission loss (TL), a basic measure of sound attenuation expressed in decibels, or more commonly as a Sound Transmission Class (STC) rating. In either case, the higher the number, the better the isolation.

Likewise you might try wall-mounted glass-fiber batts and/or geometrically sculpted open-cell foam panels concealed behind framed fabric or wall upholstery. Unfortunately, porous absorbers, so efficient at soaking up stray sound within a room, are relatively inefficient at keeping sound out. Fortifying the walls through which the offending sounds pass by double-layering your room's gypsum board walls will improve the transmission loss by only a few (probably inaudible) decibels at best.

Even sound-room walls of solid concrete some two feet thick produce a TL of only 58 decibels. (This means that sound at 88 decibels on the outside of the wall will be knocked down to 30 db on the inside). Happily if you are contemplating remodeling your listening room or building a new one, there's a better strategy for really noisy environments routinely employed by acousticians. Referred to as split or compound construction, it employs two separate wall partitions separated by an inch or more of dead air space. Each wall is constructed of conventional 2 x 4 wood-stud framing sheathed with two layers of half-inch gypsum board on one side and filled with normal thermal insulation batts. The resulting structure has an STC rating superior to that of two-foot thick concrete. The trick in compound partitions is to make sure they stay as structurally isolated from each other as possible. Any rigid tie from one partition to the other, such as a simple piece of wood or common water pipe, tends to "short out" the compound structure, converting it to a single rigid structure in which sound impinging on the outside wall surface is directly communicated to the inside wall surface, leaving us right back where we started.

The staggered-stud wall is a cousin of the double-partition wall telescoped to economize on space and materials. If there simply isn't the real estate for a double or staggered-stud wall, a second layer of gypsum board can be put close to a standard wall, but with the new layer attached to a resilient metal channel to isolate or mechanically decouple it from the rest of the wall. Finally, if there's lots of real estate, nothing beats a buffer zone or "air-lock corridor" between your A/V space and the rest of the world. Often not all walls need to be treated, but don't overlook the ceiling or floor if they are sources of external noise.

WINDOW TO A WORLD.....OF NOISE?

A structure with a high TL in most places, but a low TL here or there, usually exhibits a total transmission loss much closer to that of the low-TL element, even if it comprises only a small percentage of the total area. Think of it as the acoustician's version of the idea that a chain is only as strong as its weakest link.

Windows and doors are nearly always the weak links in the defense against the intrusion of external noise. Double panes, typically separated by a quarter-inch of air space or less, are often specified in the hopes that they'll increase transmission loss. But unless separated by an inch or more, a double-pane window or door will offer about the same TL value (or STC rating) as a single pane because the air in the narrow space between panes acts as a piston. Getting a high TL or STC rating from a window takes using:

- two panes of dissimilar thickness (one, say, 1/4-inch thick, the other 1/8-inch)

- a lossy edge mounting (1 and 1/4-inch soft neoprene strips work fine)

- a big air gap between the panes (preferably 4 inches)

- some absorption lining in the internal top, bottom and sides

- lots of care and testing to ensure that, as in compound walls, there is no rigid pathway between the partitions that would short-circuit their separation.

Louvered or hollow-core doors have poor STC ratings; solid-core wood doors and hollow-core metal doors filled with fiber perform considerably better. But it is usually the leak that does in the door. Consider that a hole or crack with a total combined area of a single square inch in a conventional gypsum board wall 8 feet high and 12 feet long — that's just one part hole per 14,000 parts wall — can transmit as much sound energy as the entire, unbroken wall! The seal, not just at the bottom of the door but all around it is, therefore, critical. High-performance acoustical doors often include "refrigerator- type" magnetic gaskets.

PANEL PULSATION

The phenomenal bass pressure generated by audiophile systems can literally flex walls of conventional half-inch sheetrock on 16-inch center-to-center studs. As they spring back and vibrate, they act like speakers — woofers, actually — relaunching a faint, out-of-tune and out-of-time copy of the original sound into the room. This background grunge is as unpredictable as it is unwelcome. Double-layering (and decreasing stud-to-stud spacing) can add damping, mass, and rigidity, decreasing the wall's tendency to "go diaphragmatic." The seams of the outer (finish) layer should be staggered relative to the inner layer and acoustical sealant used around the perimeter.

Gypsum board is not the only source of panel-type misbehavior in the residential A/V room. But as we shall see, thin panels can also control low frequency problems — including those dreaded, droning standing waves — through something called "diaphragmatic absorption."

REFLECTIONS

Sounds arrive at a listener's ears from many directions: from the sources themselves (the speakers) and from walls and objects that reflect sound toward the listener, much as mirrors reflect light. Because reflected sounds must travel further, they arrive at the listener after the direct sound with an altered frequency response and loudness level. The brain interprets these reflections differently, depending on which direction they come from, on how much later they arrive, how they're tonally changed, and how much louder or softer they are. (Curiously, reflected sounds can sometimes be louder than the direct sound if they take two or more paths to the listener — say from the ceiling, floor, and a side wall — and if the path lengths are the same so that they are additive.

A reflected sound that follows the direct sound by less than about one-fiftieth of a second is perceptually fused with the direct sound, i.e., the brain generally cannot distinguish the two as separate acoustic events. But uncontrolled, strong and very early reflections (0 to 20 msec) make a mess of perceived tonal quality and wreak havoc with stereo imaging. Reflections arriving somewhat later are interpreted as room ambience. Reflections trailing the direct sound by more than about one-fifteenth of a second can be heard as discrete echoes or more likely as reverberation. These echoes can be particularly offensive if the room concentrates or focuses such sound. Concave room features in general, such as bay windows, are frequent culprits and should be avoided if high-quality acoustic results are intended.

Getting an ambiophonic playback system to deliver the goods in a home concert hall or media room requires the elimination of as many room-generated reflections as possible. Room surfaces have three primary acoustical properties — absorption, (a complex form of reflection) diffusion, and reflection — but only absorption is of real use in the cause of eliminating audible room reflections at the listening position.

Couches, carpets, cabinets, bookcases and other furnishings all contribute to a room's reflection patterns, albeit usually in unplanned and acoustically erratic ways. For example, carpeting on a concrete or hardwood floor soaks up a fair amount of treble energy, but allows bass to bounce right back into the room. Large closed glass windows typically reflect middle and high frequencies back into the room, but let bass pass right through. A bookcase might absorb highs, scatter (diffuse) mids, and ignore the bass altogether. Thus, a room for ambiophonic listening must be treated and decorated with real reduction of reflections as the top priority.

SPLAYED WALLS

If building a new listening room or remodeling an existing room, it is possible to splay both of the side walls and the front and rear walls. The walls should lean outward as they increase in height at an angle of 5 degrees or more. The conventional wisdom has been that eliminating parallel surfaces is not worthwhile since the behavior of such a room in the bass frequency region is unpredictable in advance. But bass standing waves are not the only problem one must find a solution to.

For upper-midrange and high-frequency sounds the soundwaves coming from floor-standing loudspeakers will be reflected as light would, in an upward direction. As these rays go from wall to wall they must go up to the ceiling before they can return to ear level. Hopefully in making this longer up-and-down trip, they will lose significant energy and also fall beyond the critical 20 millisecond early reflection time zone. This is essentially a benign form of diffusion, which avoids the listening position.

In general, splaying the walls can make the absorption treatment of the walls and floor a little less critical.

SOAKING IT UP

Absorbers are devices designed to soak up sound. Most absorbers work by converting acoustical energy into thermal energy. Typically, they do this by forcing sound waves through a dense maze of small fibers that rub together to produce friction and heat. Carpet, soft furnishings, drapes and even clothing can provide useful absorption in the treble and uppermidrange, where you'll find female vocals, violins, trumpets, flutes, cymbals, squealing tires, chirping birds and other high-pitched sounds.

Acousticians refer to special sound-soaking materials like fiberglass batts as frictional absorbers, or, more colloquially, "fuzz." Generally, the thicker and denser the fuzz, the more effectively it traps sound. A dense, two-inch-thick fiberglass panel mounted directly on a wall absorbs nearly 100 percent of sound incident upon it in the range from 500 Hz (about one octave above middle C on the piano) up to 20,000 Hz, the upper limit of human hearing. To absorb much energy below 500 Hz requires a significantly thicker panel, usually 4 inches, or an air gap of a foot or two between the panel and the wall. Either way, using fuzz to soak up the lower midrange and bass requires considerable space.

REVERBERATION TIME

The amount of absorption that should be placed in a room varies according to the room's size and its primary function. All things being equal, a big room sounds more live than a small one, requiring more absorption to bring it down to the same level of acoustical "deadness." This quality is expressed as reverberation time: the amount of time it takes for sound in a room to drop 60 decibels in level from the moment the source stops producing sound. The shorter the reverberation time, or T60 as it is called, the deader the room sounds.

In general, a dedicated ambiophonic listening room should be quite dead with a reverb time of .2 seconds or less.

Because it is derived by averaging the time it takes sound to decay by 60 decibels across a broad segment of the audible spectrum, describing a room with a single reverb time figure is often as misleading as it is glib. A poorly designed room might boast a textbook-perfect average T60, yet sound disjointed and unpleasant because some frequencies die out quickly while others linger on and on. Ideally the T60 in any one-third-octave band between 250 and 4,000 Hz should not deviate from the average T60 by more than 25 percent. Translated into frequency-response terms familiar to audiophiles, this ensures that the room's reverberant sound energy is flat within about a decibel or so throughout the most sensitive range of human hearing.

One challenge lies in controlling reverberation in the bass frequencies, where T60 figures might easily be triple or quadruple that in the midrange. As Dr. Floyd Toole, past president of the Audio Engineering Society, has remarked, "At low frequencies, adequate sound absorption may be hard to find. At high frequencies it can hardly be avoided." If left unaddressed, the lack of low-frequency absorption can create an annoying unevenness in the reverberation character of a home theater, media room, or ambiophonic home concert hall.

FLEX EFFECTS

As I mentioned earlier, controlling the bass frequencies requires a lot of real estate. If space is at a premium, there is an alternative called the panel or diaphragmatic absorber. Here's the plan: Construct a 4-foot by 8-foot frame out of 2 x 8 lumber; add a single center brace (stud); attach a quarter-inch thick, 4 x 8 sheet of plywood to one side; suspend a sheet of fiberglass in the resulting internal cavity, using a wire screen between the fiberglass and plywood to prevent absorption material from touching the board. Hang the whole affair on the wall. The flexing of the panel will soak up well over 75 percent of the energy of a 70 Hz bass-drum thump hitting it, a feat that could take several feet of fuzz to equal. By varying the thickness and composition of the panel element, the depth of the cavity and the stuffing material, you can tame most any rogue bass notes that might blight your room.

Windows and even gypsum wallboard act as (normally unintentional) diaphragmatic absorbers because of their large surface areas and ability to flex. Because their efficacy as acoustical control surfaces is often unpredictable, glass and gypsum board are not often counted on for absorption in high-end A/V environments. "Decommissioning" these materials as absorbers involves making them less prone to flexing; double- or triple-layer gypsum board as well as thicker glass and smaller panes generally do the trick. As with nearly everything involved in acoustics, however, getting it right comes down to mastering the minutiae of design, installation and testing. Even the most powerful absorber becomes ineffective if it is poorly placed. Frictional types work best in some places (e.g., where air velocity is highest), while diaphragmatic designs are more effective in others (e.g., where air pressure is at its peak). For example, siting a diaphragmatic device in a pressure null — a zone of very little air-pressure fluctuation at the frequencies to be absorbed — turns a promising solution into a waste of lumber. Bass pressure tends to build up at the intersection of room boundaries, making corners especially fruitful hunting grounds for excess bass.

Tacking up 100 square feet of fuzz on each side wall yields the same absorptive value and produces the same T60 as moving the fuzz to the front and rear walls. However, the quality of sound you hear, even the intelligibility of music and dialog, could differ dramatically. In residential venues, absorption is usually best deployed on the ceiling and the front portions of the side walls, where they prevent sound from the main (front) speakers from bouncing into the listening area a split second after the arrival of the direct, speaker-to-listener sound. If left untreated, these reflective surfaces allow strong early reflections to disrupt tonal balance and imaging, and scramble the often subtle aural cues that give music and soundtracks their texture and life. However, if feasible, the ambiophonic effect can be improved by treating as many surfaces as one can bear to treat. As with most things in life, compromises may be necessary. Remember, even if your listening room is not ambiophonically perfect, neither are most concert halls.

BASS BOOMBUSTERS

Even a room that is painstakingly dimensioned to provide the smoothest possible low-frequency response will seem bass-boomy in some places, weak in others and about right somewhere else. Whether they are coaxed from genuine musical instruments, captured from environmental sources like thunder, or manufactured as special effects in a studio, all bass sounds have such long wavelengths that the resulting peaks and valleys in air pressure are clearly noticeable. You know this if you've ever walked around a room while a deep, steady note from, say, a pipe organ was playing.

The boom tends to get worse at the intersection of room

boundaries, like corners, where sound waves fold back into the room, creating a "puddle" of sound that can be as much as eight times stronger than elsewhere. In general, a bass absorber placed in a corner proves much more effective than one placed anywhere else. Other strategic spots can be discovered by computer modeling or more time-consuming trial and error.

Several companies market corner-type contraptions to lower the boom of bass buildup. Two off-the-shelf models in widespread use are the Tube Trap™ from Acoustic Sciences Corp. (ASC) of Eugene, Oregon, and the Korner Killer™ from RPG Inc. of Upper Marlboro, Maryland. Tube Traps are quarter, half, or full cylindrical constructs available in a variety of cloth-covered sizes. These simple devices feature one-inch-thick fiberglass wrapped around an interior air cavity. It takes bass sound waves a certain amount of time to work themselves through the fiberglass and into the inner cavity; this delayed equalization of air pressure is at the heart of controlling bass buildup. The bigger the diameter of the cylinder, the lower the absorption frequency. At 16 inches in diameter, the largest Tube Trap provides useful absorption down into the deep bass range.

RPG's Korner Killer works on the same frictional absorption and trapped-air principle, but it is triangular to fit tightly in the corner.

BASS BEHAVIOR

One of the most universally vexing problems of the home audio experience is the fact that residentially sized rooms give erratic support to low-frequency sounds that underpin orchestral, rock and especially modern soundtrack recordings. The perfectly recorded stroke on a drum or roar of a jet can be reproduced as a muddy drone or a wimpy whimper, depending on what the room — not the audio system — does with it.

If a bass note could talk, the first thing it would say is

"Rooms; can't live with 'em, can't live without 'em." When a particular bass note's wavelength precisely fits a major room dimension (i.e., its height, length, or width), the note is strongly reinforced in a phenomenon called a standing wave, which we readily perceive as a strong resonant boom. The room also reinforces certain higher notes that have wavelengths that are mathematical cousins of the main resonant frequency.

One of the first steps toward better sound is to design the room so that the notes that are reinforced by the room's major dimensions — its axial modes — are spread across the broadest possible range of bass notes.

Achieving smooth support for the full bass spectrum begins with getting the room's geometry and dimensions right. There are dozens of tried and true ratios: 1.0 (height) to 1.4 (width) to 1.9 (length) has achieved some celebrity, so let's put it to work by hypothetically constructing a rectangular listening room or mini-home theater with axial dimensions of 10' x 14' x 19'. Note that if we stretch the length by a single foot, we will dramatically degrade bass quality because the frequency that "fits" or resonates in the 10-foot distance between the floor and ceiling (about 56 Hz) now also mathematically fits in the room's 20-foot length. The resulting pile-up, or coincidence, of two major resonances — one from floor to ceiling and the other along the length of the room —at the same frequency will produce an annoyingly boomy response physicists call a modal degeneracy, one of those rare technical terms that carries a rich and useful connotation. Interestingly, designers of musical instruments battle the same physical problem. A strongly resonant tone on a violin or other instruments is called a "wolf note," presumably because it howls, while tones that are not reinforced are said to have a "sheepish" quality.

Also in residential construction, walls are usually not perfectly square and they flex, producing a vibration of the wall itself after the stimulus (the original sound produced

AMBIOPHONICS

by the speaker) has ceased. The resulting "overhang" imparts a muddy or fat quality to the bass, no matter how "tight" the response of the woofer or subwoofer itself. Special care in wall and ceiling construction can dramatically reduce this problem in gypsum-board walls and ceilings as well as wood-construction subfloors. This is fine for new construction, but in existing rooms the use of the bass absorption techniques already described above can tame the worst standing-wave problems even if a room is a perfect cube. RPG claims that a combination of one wall and one ceiling mounted pillobaffle in a corner can absorb 10 Sabines at 63 Hz.

CHAPTER 6
EARLY REFLECTION AND REVERBERATION SYNTHESIS

Once one decides that the audiophile approach of two-speaker reproduction of stereophonic recordings in an untreated room is never going to produce a "you-are-in-a-concert-hall" experience, there are only two ways to go if ordinary stereo records remain the program source. One is to pick a fine concert hall, construct an exact replica of it, and put two loudspeakers on the stage. That this technique would work was demonstrated conclusively several times in Carnegie Hall and Carnegie Recital Hall in the 1950s by Gilbert Briggs of Wharfedale Loudspeakers, and most notably by Ed Vilchur, the founder of Acoustic Research.

I personally attended live-versus-recorded presentations by both these gentlemen at Carnegie Hall, and not only could I not tell when the live musicians ceased playing and the recording took over, but almost no one else in the sold-out house could either, judging from the gasps and buzz in the audience when the string players finally put down their bows and the music played on. The fact that such an illusion could be created with low-powered vacuum-tube amplifiers and excellent but still relatively primitive loudspeakers should have tipped us off to the fact that ambience is essentially everything, and equipment relatively insignificant.

ACOUSTIC CARNEGIE HALL CONSTRUCTION ALTERNATIVE

As discussed in this book, it is now possible to construct a smallish room that would closely mimic the ambience of Carnegie Hall, at least in the listening area. The use of modern diffusers, absorbers, and ceiling and floor treatments could produce the reverberation time, reverberent-field frequency response and even the early reflections of any good concert hall. It would then be possible to play recordings in such a room to excellent effect. The advantages of this approach include the fact that such a room would also be excellent for live music as well.

The disadvantages of this approach for the reproduction of recorded music are several. The costs of designing, constructing and tuning such a room are beyond the reach of those of us not direct descendants of Andrew Carnegie. A major technical problem is that recordings contain ambient information that would conflict with the fixed room parameters. Both Briggs and Vilchur used their own recordings, made without significant recording-site coloration. One would also lose the flexibility of being in other acoustic settings, such as churches or recital halls.

Finally, the problem of stereo signal crosstalk would remain for closeup center listening positions. In the Briggs Carnegie Hall demonstration, (which, I believe, used mono recordings), most observers were exposed mainly to the reverberant field, and used their visual senses as a substitute for any missing or weak directional sound cues. Let us consider a more practical electronic approach.

ELIMINATION OF ROOM ACOUSTICS ENTIRELY

If we assume that we can electronically generate a close replica of Carnegie Hall, then it logically follows that we want and need no acoustic contribution from the listening room whatsoever. How to create such a non-interfering home-listening environment at reasonable cost is the subject of another chapter. But if we can create a room that has no relevant acoustic personality of its own, we can then, in principle, create any kind of acoustical signature within this space. We have to be smart enough to invent a signal processor that can generate the field we want. Once we eliminate virtually all listening-room reflections that are audible and that could possibly contradict the space we wish to listen in, there is no reasonable alternative to using a special-purpose computer to generate the early reflections and reverberation trains that thankfully the room no longer provides.

CHARACTERISTICS OF THE AMBIENT FIELD

Basically the only things you can do to a sound wave launched in an enclosed space are attenuate it or change its direction. Absorption is a form of extreme attenuation. But sound loses intensity merely by traveling a distance through air. A characteristic of attenuation is that it is almost always frequency sensitive, with higher frequencies usually attenuating more than lower frequencies, both in air, with distance, or in a sound-absorbing material. Sound changes direction whenever it encounters an obstruction — usually by reflecting, as light does (specular reflection), or by diffracting, which is a process by which sound waves sort of ooze around obstacles. As in attenuation, reflection and diffraction are frequency sensitive, with higher frequencies usually being easier to steer or control.

Thus every space, but especially a concert hall, can be described acoustically in terms of its attenuation characteristics and its three-dimensional reflectivity as a function of frequency, sound-source position and listener-seat location. The problem is then to either measure these functions in real halls and imitate them as closely as possible, or design a pleasing but entirely new hall in software that does not exist physically.

EARLY REFLECTION PARAMETERS

To produce a realistic group of early reflections, a computer or digital signal processor needs to generate and vary the following parameters separately for left and right signals:

- the delay between the direct sound and the arrival of its first reflection

- the delay of the second and subsequent early reflections and their density

- the frequency response of these discrete early reflections

- the initial amplitude and rate of amplitude loss for the subsequent reflections of these very early reflections, and

- the source of each reflection: front, side, rear, left or right.

Normally these parameters are measured in real halls, churches and opera houses, and then stored in memory. If the stored reflection patterns are not pleasing, then they can always be modified to taste. Tweaking such parameters can be a lifetime occupation, as it is with some real concert halls that are forever being tinkered with.

The parameters listed above determine how big the hall is, what its shape is (such as rectangular, fan or low-ceilinged), and how large the proscenium is.

REVERBERATION FIELD PARAMETERS

After the early reflections become so dense and weakened that the ear is no longer sensitive to their individual arrival times, the reverberant characteristics of the space become evident. The reverberation parameters that need to be simulated separately for left and right signals include:

1. reverberation decay envelope for high frequencies
2. reverberation decay envelope for low frequencies
3. frequency responses of the front, side and rear reverberation tails with time
4. density of the reverberant field
5. sources of the reverberant sound: front, sides, left-rear, right-rear, and overhead

If early reflections persist for a relatively long time before the reverberant field begins, then the space will be perceived as live and possibly large. If the reverberation time is long, then the hall will seem live or, if very long, cathedral-like. High-frequency rolloff in the reverberant field also makes the hall seem larger. The directional distribution of the reflections and the reverberant echoes help listeners determine the shape of the space and their position in it.

Again, rather than attempt to program all this from theoretical scratch, it is practical to measure several good existing halls and store the results. The Japanese, and JVC in particular, had been doing just that until the explosion of surround-sound demand forced a shift in their priorities. The JVC XP-A1010 Digital Acoustics Processor has stored within its memory the key parameters of fifteen actual listening halls, including six symphony halls of various shapes and sizes, an opera house, a recital hall, a church, a cathedral, two jazz clubs, a gymnasium, a pavilion and a stadium. Fortunately, some more recent surround-sound processors also include ambience generators that are quite suitable for ambiophonic use. The Lexicon CP-3 can generate very realistic early reflections or reverberation fields, but not both at the same time. Therefore, one has to use one CP-3 for side and rear reverb and a second CP-3 for the 55 - degree front ambience signal generation. Even a third CP-3 would be required if one wanted to use their panorama mode to eliminate front crosstalk without the barrier. We prefer to use the JVC for both front proscenium and rear reverberant sounds and one CP-3 for side ambience or reverb only.

We hope that by the time you read this, as a result of the popularizing of ambiophonics, simulators will again be produced by the Japanese, particularly Yamaha, JVC and Sony, and that the American firms of Fosgate and Lexicon will take simulation as seriously as they now take surround sound and will produce simulators with enough processing power to produce a complete ambiophonic field in one box.

PC programs that enable concert hall designers to simulate halls and predict the sound characteristics before they are built have been perfected to a remarkable degree. The latest programs and computer signal processors are now powerful enough to do this using real music input. A hall designer using near-field loudspeakers listens to how the sound changes as he shifts his imaginary seat in the hall, or changes the size of the hall or adds a diffusion cloud, etc. Such programs could easily be adapted for home ambio-

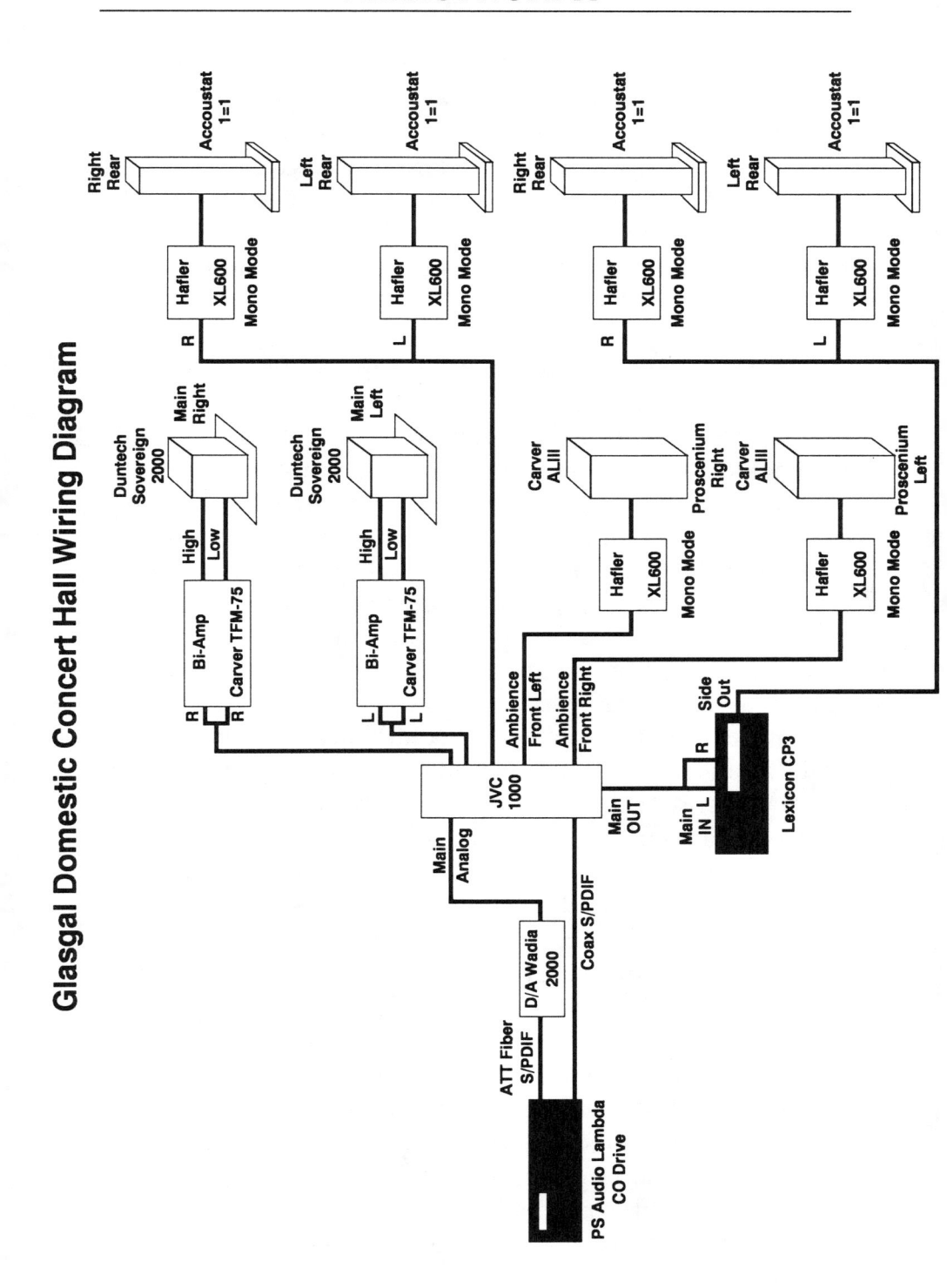

AMBIOPHONICS

Glasgal Domestic Concert Hall Wiring Diagram

85

phonic use if there were sufficient demand. Such niceties as remote control and the additional ambient digital or analog outputs would need to be added. This PC could also be controlled by the CD itself, if a standard format could be agreed upon. The PC is becoming so much a part of our daily existence that it cannot be too much longer before it also becomes an indispensible part of the home listening room.

ADJUSTING AMBIENCE PARAMETERS

To play a recording ambiophonically, one first consults the booklet or jacket to see what acoustic space it was recorded in. Was it a studio, a church, a concert hall, an opera house, a recital hall, a theater, etc? This is necessary because most recordings include recorded hall ambience that will, unfortunately, come from the front main speakers. To achieve a realistic sound field, it is necessary to match the PC/JVC/Lexicon-generated hall sounds to the recorded hall sounds as closely as possible.

You can do this quickly with a little practice by listening to the main front channels without ambience and estimating the reverberation time of the hall, which in most concert halls or opera houses is from one-and-one-half to three seconds, then estimating other hall characteristics, such as liveness. You then select the stored hall parameters in the PC, JVC or Lexicons that best match your assumptions, or custom program your guesses directly, bringing up the ambience channel volumes one at a time to the levels that sound most realistic.

The JVC does include rather effective logic, which compensates for the fact that recorded hall reverberation is being re-reverberated. But for this computer process to work properly you must tell the JVC what the approximate reverberation time of the recording is. If everything is remotely controllable from the listening position, this tun-

AMBIOPHONICS

ing process is convenient and becomes instinctive after a while; it usually takes less than a minute. Compulsive tweakers could, of course, make ambience parameter adjustment their life's work, since there are some twenty-odd volume, delay, hall type, decay and frequency response parameters that can be independently varied in many steps for each of six outputs.

The saving grace which prevents tweak insanity is that once the generated ambience sounds real and reasonably matches the recording, it can still be improved — but real is real. I found that minor shifts in JVC proscenium ambience, Lexicon side ambience or JVC rear reverberation parameters changed only the hall shape, size, liveness, and my perceived position in the hall. Someday audiophile recordings will either be made without significant recorded hall sound, or the hall parameters will be printed on the label — or even stored digitally on the CD itself for automatic electronic control of ambience synthesizers.

As part of our research process, we listened to hundreds of recordings, both LP and CD. To paraphrase Will Rogers, we haven't met a classical recording (jazz is too easy) we couldn't work wonders with. The most exciting discovery was that monophonic LP or CD recordings, even from the twenties, can be made to sound exceptionally realistic in an ambiophonic room. The reason for this seems to be that many early mono recordings have very little recorded reverberation, making it easier to create a realistic sound field to place them into. Also, the absence of a stereo effect in the presence of well generated hall ambience tells the ear/brain system that the source is distant. Thus, for large mono ensemble sound sources the listener appears to be in the balcony of a large hall — but balcony or not, real is real.

Needle scratch or frequency-response aberrations become minor distractions, and Caruso, Toscanini or Lauritz Melchior never sounded so thrilling or three-dimensional before — and the Caruso recordings are over seventifive years old! Another factor is the wall. When monophonic sound sources are listened to through spaced loudspeakers, a comb filter in the upper midrange occurs, caused by alternating signal cancellation and reinforcement, as direct and slightly delayed (.7 milliseconds, approximately) versions of the same sound reach each ear. The wall eliminates this effect and the results are a revelation. Kathleen Ferrier, in particular, has to be heard under these conditions to be believed.

MEASURING REAL CONCERT HALL SOUND FIELDS

A group of researchers in 1987 at the Victor Company of Japan (JVC), headed by Yoshio Yamazaki and including Hideki Tachibana, Masayuki Morimato, Yoshio Hirasawa, and Junichi Maekawa developed what they called a Symmetrical Six-Point Sound Field Analysis Method for measuring the acoustic characteristics of a concert hall. In their measurement method, an array of six microphones is placed at a good seat in the hall and a very loud sound impulse is launched from the stage, using a starter's pistol. All six microphones are omnidirectional, and are arranged in three pairs. The microphones in each pair are spaced about six inches apart, and thus mimic the average human head size. One pair of microphones straddles the mounting pole horizontally, left to right, one mounts front to back in the same plane and one pair mounts up and down. The center points (origin) of each microphone pair are coincident.

The impulse that each of these microphones hears then goes to a computer which produces a list of all the discrete reflections detected by the array, including their time of arrival, their amplitude and their direction. That such an array can detect all this information is not too hard to see. For

AMBIOPHONICS

instance, any impulse coming from center stage will hit the vertical pair of microphones and the pair of horizontal microphones parallel to the stage essentially simultaneously. The front-to-back pair will experience the maximum possible front-to-back delay of about .4 milliseconds. Thus when the computer detects such a situation it records that a front center-originating reflection has been received. Likewise a direct impulse from overhead will only produce a time delay in the vertical pair of microphones and a reflection from the side will only show delay in the left-to-right pair. No matter what angle a reflection arrives from, its amplitude and direction can be computed and stored.

In a real concert hall many reflections may be arriving simultaneously, so how did the gentlemen from Japan sort them out? First, each reflection of the impulse generates a signal in all six microphones. All six signals attributable to a single source will have essentially the same peak amplitude since the microphones are so close together. Thus any unequal peaks indicate a multiple reflection. Second, the time it takes for a sound to go from one microphone of a pair past the mounting pole, and from the mounting pole (or origin) to the other microphone of the same pair are always equal. Thus all three microphone pairs should record peaks that are symmetrical in time around the same origin, but with three different spacings. Thus unequal delay to and from the origin indicates an impulse collision. Finally, the ratios of these three delays define the angle to the sound source, and it happens that for such an orthogonal array, and for every impulse impinging on the array, the sum of the three cosines squared of the angle of the impulse to each axis will add up to one.

These three characteristics of the impulses detected by the microphone array represent three simultaneous equations which, when solved, allow the computer to distinguish between two and even three simultaneous or very closely arriving reflections. Since this measuring technique is relatively portable, the JVC team was able to make accurate measurements of halls like the large and small Concertgebouw of Amsterdam, the Alte Oper in Frankfurt, the Beethovenhalle in Bonn, the Philharmoniehalle in Munich, the Staatsoper in Vienna and the Koln Cathedral.

Unfortunately all this research was ahead of its time, and has not been able to establish a market yet. But this hall ambience measurement technique is so portable that every recording engineer could have such a microphone array and PC computer at any recording session. The engineer could then pick the best listening seat for the array, fire the pistol, measure the hall response and later, enter the stored results directly onto the CD for later loading into the home ambience generator, which in the future, is also likely to be a standard PC. For now such data could be gathered and included in the CD program notes.

SOME NOISE IS GOOD NOISE

When you are in a concert hall and the music stops, you are still in a concert hall. Even with your eyes closed you can sense a sort of ambient ambience, a murmur or acoustic dither that even without an audience present tells you what kind of acoustic space you occupy. By contrast, in the ambiophonic hall, when the music stops, you are abruptly transported from a lively exciting space to a rather dead, sound-treated listening room. For CDs with many silent bands between short selections, we have found this effect is somewhat disconcerting.

One solution, if you are similarly afflicted, is to actually go to a church or large hall while it is empty, record about two hours' worth of background noise on a cassette. This cassette can then be run continuously on its own amplifier and speaker at very low level to fill in the silent gaps.

AMBIOPHONICS

Another approach, which is the one we use, is to play a pink noise test CD through a spare subwoofer. Like sensurround™ in movie theaters, the source of this building rumble cannot be located.

CHAPTER 7
FUTURE AMBIOPHONIC ENHANCEMENTS AND SURROUND SOUND ROUNDUP

AMBIOPHONIC RECORDING TECHNIQUES

Most of the present stereo recordings available were not made with ambiophonic playback systems in mind. The problem with such recordings, from the ambiophonic point of view, is that they contain both early hall reflections and hall reverberation mixed with the left and right direct-sound and early proscenium reflections. Thus, when these signals are used to generate reflections and reverb in the home concert hall, we must be generating reflections of these recorded reflections and reverberation of this recorded reverberation. But even worse is that most of this superfluous hall ambience is coming at the listener from the main front speakers, which is totally unrealistic.

Anechoic recordings, which have no recorded reflections or reverb at all work well but not as well as you might think. This is because in concert halls there are actually early reflections and some proscenium reverberation that hit the ears from the front. Thus the ideal ambiophonic recording is one using closely spaced, directional microphones where the microphones face forward and together can only pick up sound from the 180-degree angle in front of them. Again, this frontal recorded ambience will be wrongly recycled by the ambience generators, but since this frontal ambience is mostly very highly correlated and is double delayed, the effect is minimal. Finally, rear hall sound that reflects off the proscenium and is picked up by the microphones is also late, attenuated, correlated and comes from the front where the ear is less sensitive to this type of error.

In any case, depending on the recording, our home concert hall may not be acoustically perfect, but even an imperfect hall is still real. One could say that the additional reverberation of the recorded front reverberation is equivalent to installing an extra layer of diffusion clouds above the stage of a concert hall or adding an alcove to a cathedral.

Recordings that include an excessive amount of long-period reverberation are difficult to optimize, but not hopeless, since the IACC of this front-arriving reverb is so high that it is somewhat swamped by the simulated reverb; at least that reverb is arriving from the proper side and rear directions. Another trick is to enlarge the space. If a recording is made in a very live recital hall, move it into a larger concert hall — or move the church choir into a cathedral — thus minimizing the effect of the shorter recorded reverberation field. We can hope that eventually some specialty record companies will produce recordings better tailored to ambiophonic reproduction.

AUTOMATIC CONTROL OF AMBIENCE SYNTHESIZERS BY CDs

If the ambiophonic method catches on, there is no reason why recording engineers could not encode the various early reflection and reverberation parameters they feel appropriate in the preamble digital code of the CD. This data could then directly control an ambience simulator built into the CD player or externally. A standard format for this would need to be established. In the meantime, if such information were included in the CD booklet this would be a big help.

It is also possible, knowing these recorded reverberation parameters, to subtract much of the recorded hall reverberation from the main front channels. This would solve the problem covered in the previous section. With computer and digital signal processing chips becoming ever faster and cheaper, the day is not far off when these capabilities will be taken for granted.

ADDING VIDEO TO AMBIOPHONICS

The problem with playing laser disks of concerts or operas in the home concert hall is that the crosstalk barrier blocks the view of the listener to the front, making placement of a conventional TV set difficult. Two recent developments have made it possible to experiment with video in the home concert hall. One is the flat-panel LCD video monitor. You can hang a 4-to-7 inch flat video monitor, like a framed miniature, right on the edge of the wall in front of you. I have done this, and the only problem I had was that I had to wear reading glasses when listening since the screen was only about 20" away.

Another approach is the virtual reality TV monitor. Several manufacturers have built LCD video monitors into computer-style virtual-reality-type goggle frames. You simply wear them like a giant pair of sunglasses and a very large TV image appears visible to one or both eyes. The image, in fact, occupies so much of the field of view that there is a slight impression that one is watching big screen projection TV. Although the picture resolution presently leaves something to be desired, this approach seems very promising. Incidentally, viewing a picture while listening considerably dulls the sensitivity of the ears to ambiophonic factors such as space and directionality. This may explain why so many viewers are so satisfied by surround sound. Perhaps surround-sound systems should be judged with the eyes closed. This goes for live concert-hall design critiques as well.

DISCRETE MULTI-CHANNEL RECORDINGS

There is absolutely no reason why eight-channel discrete ambiophonic recordings could not be made. This would eliminate the problem of re-reflecting reflections and re-reverberating reverberation. But everything else, including the crosstalk elimination wall, room treatment, proper speaker selection, speaker placement and the fixed listening position would remain. It is, therefore, quite unlikely that such a format will become a recording standard, since ambiophiles will never add up to a mass market. One could also argue that having to use an ambience synthesizer is not a serious technical disadvantage, since it allows hall seasoning to your own taste and is probably more cost-effective than paying a premium for each and every eight-channel discrete recording purchased.

The new AC-3 5.1 discrete, six channel, digital, recording standard already adopted for laser discs, provides left, right and center front full range signals, two independent rear channels and a subwoofer channel. Assuming that the compression algorithm used in the system proves to be robust enough to reproduce music with high fidelity, AC-3 is a viable vehicle to deliver concert-hall sound to an ambiophonic reproducing system. Some changes in the definitions of the signals are required when shifting the format from video audio to ambiophonic audio. Thus the two rear channels are likely best used for left and right 55-degree early reflections and the center channel used for a mono-rear reverberation signal. The sub-woofer channel can remain as is or be ignored. While not ideal, there is nothing to prevent one from synthesizing additional signals from those present and fleshing out the soundfield. In the end, however, the difficulty of still needing a barrier, having two sets of speakers in different positions in one room or two rooms, one for surround video and one for ambiophile audio, is too great for this to ever be practical or find a mass market.

SURROUND SOUND ABOUNDS

Where does surround sound fit into the ambiophonic scheme of things? There are two major differences between the ambiophonic concept and the usual surround-sound systems used by the TV and movie industries. In current surround sound, all surround signals are extracted directly from the two matrixed signals. Signal processors monitor these two input signals and decide, on the basis of their short-term relative amplitude and phase, what part of each input should be steered to which of the loudspeakers encircling the listening position. Unlike the ambiophonic method, which uses a synthesizer to locally create the signals for the surround ambience, the surround-sound system, if used to reproduce concert-hall music, attempts to extract the reverberant field sounds inherent in the recording and to steer them to the appropriate speakers, being careful not to impair front-channel separation or to allow front-channel leakage to the ambient speakers.

The second difference is that the "wishful thinking" function of surround sound is to provide the listener with the possibility of hearing sound images from any direction. Surround sound puts the actor, musician, or other sound source into the viewing room. Sound effects or dialog can appear to come from anywhere in the theater, but as exciting as this may be, it is not the same as being there, and there is no serious attempt to move the viewer to the scene by

AMBIOPHONICS

changing the acoustic character of the viewing room. In the ambiophonic system, since a CD can run for seventy minutes or more, it is practical to set up a realistic ambient field and move the listener into it for the duration. In the case of a movie, the ambience generator settings would have to be changed automatically with every scene, and this would require an encoding standard for which there is no existing or probable market. Furthermore, such an ambience-varying system would only work well in an ambiophonically treated viewing room.

Comparing surround sound to ambiophonics is not completely logical. Ambiophonic methods assume the listener has been moved to a concert hall and therefore expects to hear direct sound coming only from the stage in front of him or her. Thus two-channel, unencoded, non-matrixed discrete recordings are completely able to produce the line image desired. The only disadvantage of this surround dimensional lack that we have noticed is that audience noises, such as applause in live recordings, always appear to be up front.

To get a really reliable 360-degree direct surround-sound effect requires either very sophisticated logic circuits to steer signals among multiple speakers surrounding the listener, based on a specific matrix encoding standard, or true discrete multichannel recordings. We have already considered the case of multichannel recordings where the extra channels are used to record hall ambience. Presumably if multichannel surround-sound CDs are made for musical audio-only purposes, they could easily be adapted so as to sound good in the ambiophile home concert hall. But playing any surround-sound music over the usual surround-sound speaker array in an untreated room can never produce a truly realistic "you-are-there" sound field. First, the crosstalk problem between the front left and right speakers remains — even with a center dialog speaker improving center sounds by swamping some of the crosstalk, but at the expense of stage width. Furthermore, almost all the critical early reflections come from the wrong directions, even the best decoders cannot really extract the reverberant field intact, and the untreated listening room distorts anything that is extracted.

However, if encoded, matrixed, multi-channel surround-sound audio or audio/video recordings of opera or concerts become commonplace, they could be accommodated in the ambiophonic home concert hall pretty much as described above for discrete multichannel recordings. For two-channel encoded sources such as Dolby Pro Logic, the decoded left-and-right front signals could be used to generate early side reflections and some rear reverberation. The decoded side ambient signals can often be left as they are, coming from the side speakers. The front center speaker should be turned down and the crosstalk barrier put in place, assuming there is no video. All this assumes that the room has been sound treated, but even a pure surround-sound system would benefit from this.

When all is said and done, a practiced ambiophile, using a synthesizer, can probably produce a more realistic, and certainly more stable, sound field for music directly from the undecoded surround-sound input signals just as they are, than from using any presently envisioned matrixed surround-sound system.

EPILOGUE

One reason for writing this book is to encourage audiophiles like the writers and readers of *Stereophile*, *The Absolute Sound*, *Audio*, and *Stereo Review* to get behind ambiophonics and encourage manufacturers and recording engineers to design products and make recordings that exploit this methodology, and thereby perfect it to the point where even Barry Willis comes to believe that you can accurately "be there" musically while still at home bodily.

Please understand that we do not claim to have reached perfection in sound reproduction — only another step in that direction. We believe that the ambiophonic approach creates an acoustic field good enough to fool the brain, but that this illusion is usually far from perfect, even if real. It may be too dead, too shrill, too hard, too soft, too noisy, too small, too big, too reverberant, or too bass shy. Your seat may be too close or too far back, and the stage may be too wide, too shallow, or too narrow. However, the fact that such factors become clearly audible, and could also apply to live performance experiences, is, to my mind, proof of an advance in the state of the reproductive art and provides a more rational framework to proceed from here to advance that art still further.

APPENDIX

LIST OF COMMERCIALLY AVAILABLE AMBIENCE SYNTHESIZERS SUITABLE FOR USE IN AMBIOPHONIC LISTENING SYSTEMS

Fosgate Audionics 3A
JVC XPA 1010
Lexicon CP-1 Plus
Lexicon CP-3 Plus
Soundstream Technologies C.2THX

BIBLIOGRAPHY

Ando, Dr. Yoichi. *Concert Hall Acoustics*. 1985. Munich: Springer Verlag.

Bock, T.M. and Keele, D.B., Jr. "The Effects of Interaural Crosstalk on Stereo Reproduction" and "Minimizing Interaural Crosstalk in Nearfield Monitoring by the Use of a Physical Barrier," presented at the Eighty-first Convention of the Audio Engineering Society, Los Angeles, California, November 12-16, 1986. Preprints 2420-A and 2420-B.

Stereophonic Techniques — An Anthology of Reprinted Articles on Stereophonic Techniques. 1986. New York: Audio Engineering Society.

INDEX

Absorbtion 39, 41, 42, 52, 57, 61-64, 68, 70-72, 74, 76
Acoustic Sciences Corp. 62, 63, 75
Ambience Generator 4, 79-91, 96
Ando, Yoicho 5, 18, 21
Audio Engineering Society 73
Autocorrelation 12, 16, 17, 18

Barrier 8, 37, 39-44, 52, 44, 60, 63, 88, 97
Binaural 13, 5, 9, 35, 42, 43, 47, 49

Carnegie Hall 79, 80
Carver, Robert 36
Comb filter 32, 34-35, 37, 48, 88
Crosstalk 29-33, 35-37, 40-42, 46-48, 51, 80, 96, 98, 99

Diaphragmatic Absorber 69, 73-74
Diffusion 54, 70-71, 84, 94
Dolby 46, 96, 99

Fiberglass 58, 66, 72-73, 75

IACC 12, 16, 18, 19, 21, 26, 33, 55, 94

JVC 46, 83-84, 86

Keele, D.B. 32
Klayman, Arnold 45-46

Lexicon 36, 46, 83, 84, 86

Matrix 4, 99
Modal Degeneracies 14
Monophonic 26-27, 43, 47, 88, 96
Multichannel Recording
MSB Technology 42, 63

Noise 13, 65-68, 90-91

Personal Computer (PC) 84, 86, 90
Phase 28, 47-49, 97
Pillobaffle 61, 77
Pinna 10, 18, 32-33, 35-36, 43-47
Polarity 12, 49-50
Polk 36

Reflection 8, 11, 16, 18-23, 28, 47, 50-52, 54, 58-59, 63, 69-70, 74, 79-91, 93-99
Reverberation, Reverb 4, 9-10, 18, 22-23, 26, 45-47, 51, 55-556, 58, 60, 63, 72-73, 79, 93-99

107

AMBIOPHONICS

RPG 61, 63, 75-77

Sabines 62, 65, 77
Shadow Caster
Splayed Walls 71
STC Rating 67-68
Surround Sound 1, 36, 46, 61, 93-99
Synthesizer 8, 54, 79-91, 96, 99

Toole, Floyd Dr. 73

Willis, Barry 1-2, 101

Video 3, 8, 61, 95, 97, 99

ORDER FORM

Fax Orders: 201-768-2947
Telephone Orders: 800-526-9261
Postal Orders: Ambiophonics Institute,
      c/o Ralph Glasgal
      151 Veterans Drive
      Northvale, NJ 07647
      201-768-8082 ext.205

Please send ____ copies of Ambiophonics to:
Name_____
Address_____
City_____ State_____ Zip_____
Country_____
Telephone_____

    Sales Tax: Please add 6% for books shipped
        to New Jersey addresses.

Price of U.S. $29.95 per book includes book rate postage by U.S. Mail. (Allow 3 to 4 weeks for delivery). Those ordering by credit card, blank cheque, or C.O.D., may request shipment at additional cost by UPS, Federal Express, Air-Mail, First Class Mail or International Carrier.

Payment:

☐ cheque

☐ credit card: ☐ VISA ☐ MasterCard ☐ AMEX

Card Number_____

Exp.date_____/_____/_____

Name on Card_____

☐ C.O.D.